THE COLOR OF NORTH

THE COLOR
OF NORTH

~ഡഡ~

THE MOLECULAR
LANGUAGE OF PROTEINS
AND THE FUTURE OF LIFE

SHAHIR S. RIZK & MAGGIE M. FINK

THE BELKNAP PRESS OF
HARVARD UNIVERSITY PRESS

CAMBRIDGE, MASSACHUSETTS
LONDON, ENGLAND

2025

First printing

Library of Congress Cataloging-in-Publication Data
Names: Rizk, Shahir Samir, 1978– author. | Fink, Maggie M., author.
Title: The color of North : the molecular language of proteins and the future of life/ Shahir S.
   Rizk, Maggie M. Fink.
Description: Cambridge, Massachusetts ; London, England : The Belknap Press of Harvard
   University Press, 2025. | Includes bibliographical references and index.
Identifiers: LCCN 2024042289 (print) | LCCN 2024042290 (ebook) |
   ISBN 9780674292581 (cloth) | ISBN 9780674300514 (pdf) | ISBN 9780674300521 (epub)
Subjects: LCSH: Proteins. | Proteins—Structure. | Protein engineering.
Classification: LCC QD431 .R57 2025 (print) | LCC QD431 (ebook) |
   DDC 572/.9—dc23/eng/20240929
LC record available at https://lccn.loc.gov/2024042289
LC ebook record available at https://lccn.loc.gov/2024042290

*For Dr. Gretchen Anderson, forever
our teacher, mentor, and friend*

CONTENTS

THE COLOR OF NORTH

PROLOGUE

WHEN PEOPLE THINK of proteins, they often think of food. It's true that proteins are a major part of our diet. But proteins are also so much more. Inside every living cell, countless proteins go about their business—the business of life. Proteins are the microscopic worker bees of the cell. They are the tiny machines that facilitate nearly all biological functions in every organism that has ever lived. They power our very existence.

When we set out to write this book, we had one goal: to share our love of proteins. Each of us had arrived at the world of these cellular acrobats in a different way. Growing up on opposite sides of the globe, we each lived through different events, experienced different traditions, and spoke a different language. But eventually one thing united us—our love for science. In 2017 we met on the campus of Indiana University South Bend and began working together on a research project. Our goal was to use a protein from bacteria to detect pollutants in soil and drinking water. We worked in a small lab in the basement of the old Northside Hall building. Under buzzing neon lights and crumbling ceiling tiles, we watched bacteria grow on Petri dishes. We set up clever experiments to extract the protein we wanted to study. We used lots of machines—some big, expensive, and loud; others old, rackety, or home-made. As we learned more about how proteins work, our passion for studying them only grew. These biomolecules were alive. They danced and wiggled. They shape-shifted and transformed.

We are scientists, but we are children at heart. Both of us became fascinated with proteins the first time we heard about them. We were struck by a childlike amazement when we saw their enigmatic structures and learned about all the things they do. In our research together, we would stare at the shape of proteins on computer screens, marveling at every twist and turn. We lived and breathed proteins, and came to see how each protein had a personality, a "style," of its own and behaved in particular ways. The proteins we studied became like friends. And sometimes, like many of our human friends, proteins surprised us with strange behaviors and we were left perplexed, even heartbroken. Working with proteins over many years, trying to understand their behaviors and gain insight into how they work, has sometimes left us frustrated, but it has also brought us immense joy. It is this joy of discovery that we most want to share.

The study of proteins spans many disciplines, most notably chemistry and biology. Proteins form through chemistry—that is, an assortment of atoms, such as carbon, oxygen, and nitrogen, assemble in specific arrangements. Then these collections of atoms come to life, and all that chemistry becomes understood through the lens of another scientific field, biology. Perhaps this is why exploring the world of proteins can be exclusive. Often, one does not get to dive deep into the world of protein structure and function until late into college, after several semesters of both chemistry and biology. Sadly, this can turn many creative minds away from learning about these amazing molecules, and leave most people unaware of the busy microscopic citizens working inside each one of our cells. With this book, we hope to make the world of proteins come alive with explanations that readers of any scientific level can grasp and everyone will appreciate.

We are scientists, but we are also storytellers. Throughout these chapters we take turns telling our stories, writing about our relationships with friends and family, moving to new places, and losing those we love. In other chapters, we collaborate to tell the stories of those who helped spark our passion for proteins. In all these stories, we explore the role of proteins in our human journey through life, from birth to death. We investigate how webs of proteins build the foundation of our most intricate cellular structures. We examine how proteins enable us to perceive our surroundings, ultimately shaping our experiences of the world. We also see how proteins give some animals incredible senses that extend beyond the confines of our own sensory experiences. We tell stories of proteins that protect fish and frogs from freezing to death, and others that help bacteria survive in boiling water. We peer into the proteins that make fireflies flicker and the ones that make jellyfish glow. We explain how some proteins protect us from danger, how others help plants and bacteria remember, and how some make up the venom that snakes and scorpions use to paralyze their prey. We describe how faulty proteins can trigger disease or even signal our demise, while others are used by bacteria and viruses to invade our bodies. But we also reveal how newly engineered proteins can bring about cures for deadly diseases, bringing life where death seemed inevitable. And we look ahead to a future where "designer proteins" can be a valuable weapon in the fight against pollution and climate change, offering hope for our planet.

We are scientists, but we are also artists, captivated by the fascinating structures of proteins. Throughout this book, our hand-drawn illustrations showcase not only the sheer variety of protein structures found in a wide range of organisms, but also how these shapes support the functions that sustain life. The images of the chemistry and 3D structures of proteins demonstrate that anyone can visualize proteins, and are a reminder of how art often enriches our understanding of science. Under each illustration, a unique ID is listed. This is the unique code for the protein structure found in the Protein Data Bank (PDB), a repository of all known protein structures built by contributions from scientists from every corner of the world. We invite you to learn more about each structure by going to the PDB website (www.rcsb.org) and typing in the unique code found under our illustrations. On that website, you will be able to learn more about each protein and even use your mouse to magnify and flip its 3D image every which way, so you can examine it closely on your own screen.

This book is an outpouring of love, a tribute to the beauty we witness and the inspiration we feel every day when we step into our labs and have the opportunity to learn more about proteins. We invite you to join us as we share what we now know about these fascinating molecules, our companions since the dawn of life that continue to shape the future of our planet.

1

ORIGIN

SHAHIR S. RIZK

USING HER STONE mortar and pestle, my grandmother crushed fresh garlic cloves with cumin, sea salt, and a drizzle of lime juice, filling the kitchen with a powerful aroma. On the small table to the side, half a dozen fish, cleaned and gutted, sat on a metal plate. She handed me the heavy stone container and instructed me to spread the paste on the fish, covering every inch, while she washed the rice. My hands would smell like garlic for the rest of the day, but I didn't mind.

Though we didn't rush this process, my grandmother and I needed to keep track of time. Soon my parents would return from work, hungry, and we would have a meal ready for them as we had done every day over the summer. Each meal took the entire day to prepare. I listened to my grandmother's instructions and observed the methodical but relaxed way she dipped the fish in flour, slapping each one between her two palms to remove the excess and slipping them into a pan of sizzling oil.

My grandmother grew up in the south of Egypt in a rural town where she learned the culinary arts of the region. After marrying my grandfather, they moved north with the flow of the Nile River, living for several years in Cairo, where my mother was born. They ultimately settled farther north in Zagazig, a small city in the ancient land of Goshen, where Joseph and the Israelites had settled thousands of years before. Zagazig was my childhood home, too, and I lived just a short walk away from my grandmother's apartment. Spending summers with her was an opportunity to learn the old ways of the world. It was a chance to grasp a piece of the past, a past that was getting pushed away by the modern world with its frozen meals, disposable culture, and digital toys. I spent my summers watching, learning, and of course, cooking.

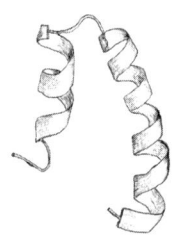

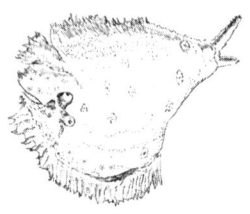

Pardaxin is a small protein secreted by the Moses sole when threatened (PDB code 2KNS). The protein acts as a powerful shark repellent.

Our seafood dishes often contained the Red Sea Moses sole, a strange fish that resembles a flounder. The eyes of the Moses sole point up from the top side of its flat body. Its mouth is tilted toward one side, and its lopsided gills give it a permanent grin. The bottom side is bare, devoid of any recognizable fish features. "It looks like a fish split in half," I said. With a skeptical smile, my grandmother recounted how, when Moses parted the Red Sea, one unlucky fish happened to be caught in the middle of the action; it too was split in half, and that's the ancestor of what we eat today. We both chuckled as she turned the burner down low under the steaming pot of rice.

The Moses sole lives near the bottom of the Red Sea, where the water is warm year-round and teeming with life, including hungry sharks. With both eyes pointing upward, the Moses sole is always watching for predators, and its dotted scales match the color and texture of the sandy sea floor, providing camouflage. But if it's not hidden well enough, and a hungry shark does spot it, the fish has another weapon: it secretes a powerful shark repellant known as pardaxin. When released from the Moses sole, pardaxin coats the inside of a shark's jaw, causing temporary paralysis and preventing the Moses sole from becoming shark dinner.

Pardaxin is a protein, and, like all proteins, it is invisible to the naked eye. It would take about one million pardaxins, strung side-to-side, to equal the width of the period at the end of this sentence. Even the most powerful microscopes struggle to get a glimpse of proteins. Scientists must use x-rays and sophisticated machines, along with a lot of computations, to determine what proteins look like. Deciphering the structure of a protein is a heavy lift, but it's worth it, because the secret to how any protein works lies in its unique structure.

When we zoom in on pardaxin, we see that it resembles two twisted spiral staircases connected by a short loop, making a "hairpin" shape that resembles a bobby pin.[1] Each of the spirals is known as a helix. Positive charges along the loops are attracted to negative charges on the surface of cells lining the inside of the shark's jaw and gills. When the Moses sole senses an approaching shark, it releases a milky cloud containing millions of pardaxin molecules. Each protein becomes attracted to the shark's jaw and gills much like a magnet jumps from your hand and onto a refrigerator. Once latched on, the pardaxin molecules act like thousands of tiny thumbtacks, poking miniscule holes in the shark's cells and causing their insides to leak out as salty ocean water rushes in. The disruption of the cell membrane and the imbalance of salts within the shark's cells temporarily paralyze the predator, giving the Moses sole enough time to escape. Having a shark repellent at the ready has helped the Moses sole survive in the Red Sea for millennia. But pardaxin is only one of thousands of proteins made by the fish, each with a specific structure and a precise function. Together, they make life at the bottom of the Red Sea possible.

Proteins are not unique to the Moses sole; they are found in every organism, from bacteria to bananas, birds to bees, algae to alligators, and mold to chimpanzees. We make our own proteins and use them for almost everything we need to survive. Each one of our 40 trillion or so cells is packed full of proteins swimming shoulder to shoulder in a solution of salty water. In fact, proteins make up about half of the molecules in our cells. Most of the other half is water, and the tiny remaining percentage is comprised of DNA, sugars, and lipid (fat) molecules.[2] If one of our cells was the size of an average American home, it would be filled with about 30 billion proteins, ranging roughly from the size of a grape to that of a watermelon. Each of these proteins carries a story of who we are and where we come from—and not just our family history, but also our evolutionary history, that is, how we are linked with everything around us. Proteins convey our origin story and potentially, the origin story of life itself.

INGREDIENTS FOR LIFE

My grandmother was born before World War I and left school after the fourth grade. She was married and already had several children when she survived the German air raids over Cairo during World War II. She didn't understand the new fads of the 1980s. To her, my pixelated computer games seemed to be from a different world. But she knew many secrets rooted in science without having learned them at a lab bench or from a book. She knew that crushing garlic released its aroma. We now know that this breaks up the cells of the garlic plant, releasing a special protein—an enzyme that drives a chemical reaction, liberating molecules with that characteristic smell. She knew that if you boil stew for several minutes and leave it covered, it will keep for days. We now know this as pasteurization: the high heat unravels and inactivates many toxin proteins, killing bacteria that would cause sickness. Without knowing anything about proteins or enzymes, my grandmother relied on ancient wisdom, wisdom that had flowed down from one generation to the next in stories, anecdotes, and food just as the Nile River had carved its way through desert rocks, transforming the sand into a narrow, lush valley of verdant farmland. And it was that ancient wisdom that I tried to soak up.

Throughout history, stories and traditions have been passed down to help future generations understand their world and thrive in it. How to crush garlic. How the Moses sole came to be. How to listen for the sound of the water reaching a boil. In this way, we inherit information from our ancestors, information as important as the genetic information we receive from our parents in the form of DNA. Like a handwritten recipe, jotted down on a stained scrap of paper and passed from generation to generation, our inherited DNA is the instruction manual for making proteins. Every gene in DNA is a code for making one protein. Slight differences in the way our proteins are constructed give us curly or straight hair, make us short or tall, and determine how much of the pigment melanin our skin and eye cells make. But beyond our eye color or hair texture, our proteins are essential to every breath, heartbeat, and blink of an eye. Proteins dictate how we interact with the world around us. They signal when it's time to wake up and when it's time to sleep. Proteins convert signals from light, scents, and tastes into a molecular language carried through a chain of hundreds of proteins that move and twist, resembling a microscopic Rube Goldberg machine. As it makes its way through our nervous system, this molecular language is translated and perceived as colors, shapes, flavors, and smells. Proteins are quite literally interpreters of our environment.

Humans have roughly twenty-five thousand different genes. Most of our genes contain the information to make a protein, or a segment of a protein, with a unique shape and a specific job.[3] Even though almost every cell in our bodies carries the same genetic material, not all twenty-five thousand proteins are made in every cell. The difference between a liver cell and a brain cell is not the genes they carry but which genes they "read" to make proteins. Liver cells will read only the parts of DNA required to make the proteins that a liver cell needs. The other genes remain dormant. The same is true for a brain cell, an immune cell, or a skin cell. Some genes can remain dormant for the entire life of a cell, while others are brought to life to create fully functioning proteins. And so, while the DNA within a cell can tell us whom the cell belongs to, the proteins the cell chooses to make tell us the type of cell it is. Each cell contains its own story, with proteins as the main characters acting out the directions of the DNA.

But proteins are not directly made from DNA. There is a go-between molecule that facilitates the translation of genetic language into functional protein language. This molecule is RNA, an essential compound in all organisms. In fact, it is RNA that assembles the building blocks of proteins, connecting them in a specific sequence based on the instructions encoded in DNA. In this manner, proteins in every cell are assembled by the action of RNA to carry out all the functions that cells need. This flow of genetic information from DNA to RNA and into proteins is the central dogma of biology. Most of our understanding of this process has come from studying genes. Similarly, scientists have mostly looked at evolution through the lens of DNA and mutations in the genetic code. But as more tools and technology have allowed proteins to be studied, questions about the origins and evolution of life are now being looked at from the point of view of proteins, the final products of DNA. By shifting our attention to proteins, to the stories of evolution, adaptation, and inheritance they hold, our story and the story of all living things become clearer.

There is no known lifeform that does not rely on proteins, so the question of how proteins became integral to life is a question about the origins of life itself. We still know little about the conditions that made life possible on our planet, but these conditions were likely

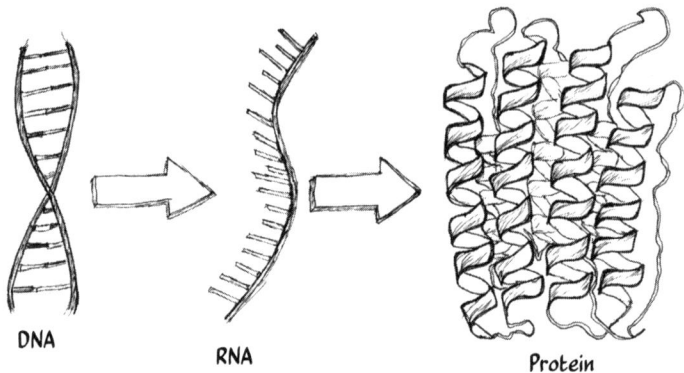

DNA

RNA

Protein

According to the central dogma of biology, genetic information flows in one direction from DNA *(left)* to RNA *(middle)* to protein *(right)*. DNA is transcribed into the RNA, which is then used to make proteins that carry out essential biological functions.

very different from today's. Geologic records paint a picture of a harsh environment with volcanic eruptions spewing toxic gases of methane and ammonia. Without the atmosphere we have today, the Earth's surface was bombarded with ultraviolet light, and intense storms covered large parts of the world. In this chaotic environment, simple, organic molecules began to come together, fusing to make more complex molecules, some of which may have become the building blocks of DNA, RNA, and proteins.

While it might seem unlikely that life could emerge in such harsh conditions, experiments conducted in the 1950s by Stanley Miller and Harold Urey at the University of Chicago suggest that this scenario was not far-fetched. Miller and Urey tried to recreate the conditions on our planet when life first began 3.5 to 4 billion years ago.[4] In a sealed container, they mixed volcanic gasses of ammonia, methane, and hydrogen above saltwater that resembled an early ocean. Then they heated the mixture, periodically zapping it with an electric spark to simulate lightning from a storm. Within one day, the color of the solution trapped in the reaction vessel began to change to a faint pink. After a week, it turned a deep red. When Miller and Urey opened the sealed container, they found that the mixture now contained five different amino acids, the building blocks of proteins. The experiment was groundbreaking, demonstrating the possibility that the molecules needed for life could have been made from basic substances likely found in a primordial Earth.

Even with the Miller-Urey experiment, many questions remain unanswered. What did the first protein look like? When and how was it made? We still don't have the tools to answer these questions. But we do know that proteins are a part of every plant, animal, and even bacterial molecular story—and that learning more about proteins can shed light on the numerous ways nature has used them to navigate the extreme ups and downs of the world and to ensure the survival of all organisms.

Proteins are incredibly versatile molecules that play a wide range of roles in living organisms. They support the basic functions that are common to all life. Every organism must eat, grow, reproduce, and excrete waste. Underlying these common functions are proteins that have not changed much across the kingdoms of life. For example, many of the proteins we use to digest our food are nearly identical to those used by bacteria, our most distant relatives. Likewise, most of the proteins that are used to duplicate DNA every time our cells divide are similar across all forms of life. Many proteins act as enzymes, facilitating the chemical reactions within living cells. These proteins synthesize and recycle a wide range of molecules and generate energy from the food we eat. Structural proteins support cells and tissues underlying the framework of skin, bones, hair, and nails. They also form the fibers of muscles and tendons, enabling us to move when our muscles contract. Transport proteins help move molecules throughout the body, delivering nutrients to cells, removing waste, and flushing out harmful toxins. Signaling proteins are involved in all cellular communication; many hormones are signaling proteins that regulate various physiological processes. Receptor proteins on our eyes, tongues, and noses help us see, taste, and smell. And immune-response proteins identify and neutralize foreign invaders such as bacteria and viruses.

Having such amazing molecular machines helps organisms adapt to their environment, survive, and proliferate. But when proteins go rogue, they can play a significant role in disease progression. One of the most devastating examples of proteins gone awry is cancer. The uncontrolled cell growth eventually leads to tumors that often spread throughout the body. Proteins called oncoproteins can fuel this growth by promoting cell division or inhibiting natural cell death. Proteins known as tumor suppressors, which normally act as a check on cell growth, can be inactivated, allowing cancer cells to proliferate unchecked. Protein disfunction plays a significant role in neurodegenerative diseases like Alzheimer's and Parkinson's, too. As proteins fail to take on the correct structure, they accumulate in the brain, where they form toxic clumps that disrupt normal cellular function.

Proteins also play a critical role in infectious diseases. Pathogens like bacteria and viruses often produce proteins to help them evade the immune system or to hijack host cells for their own reproduction. The spike protein on the surface of the SARS-CoV-2 virus, for example, allows it to enter human cells and cause COVID-19. Once inside our cells, the virus can make copies of itself using the cell's own protein-making machinery. Without the action of the proteins produced by our immune cells, the virus spreads without control, causing disease and even death. Understanding the role of proteins in disease progression is crucial for developing effective treatments.

Just as so many proteins have mostly remained unchanged throughout evolution, some have come to possess truly incredible properties that defy our own imaginations. Proteins help bacteria survive inside nuclear reactors and protect microorganisms from the immense pressure and scorching temperatures found near deep-sea vents miles below the ocean's surface. They keep birds in sync with the magnetic field of the Earth as they migrate each year and give fireflies their ghostly glow on warm summer evenings. And for a select few organisms that live in cold climates, proteins prevent them from being frozen solid. By looking closely at these miracle-performing proteins, we can start to build an understanding of how all proteins, even those that have dramatically different tasks to perform, evolve.

SLEEPING BEAR

Each summer, I looked forward to my cousins' arrival, lazy afternoons cooking with my grandmother, and much-anticipated trips to the coast of the Red Sea. Summer was special. Twenty or so cousins and all of our parents would cram into a house, a stone's throw from the beach. We would swim from sunrise to sunset and gorge on watermelon and pretzels, fried fish, spicy shrimp, and ice cream. We would bake under a perpetual sun that turned our olive skin even darker, wincing later as our parents rubbed aloe on our tender shoulders. We would splash in the warm waters and compete to see who could stay underwater the longest. The only thing I knew at the time was warmth. In fact, the first time I saw snow, I was fifteen years old, right after my family moved to

America. Our first winter in the United States was one of the coldest on record, and we happened to spend it in one of the coldest places in the country: Sioux Falls, South Dakota. A few days after we landed, the day before Halloween, it began snowing, and the snow did not completely melt until the end of May. By November, temperatures had dropped below $0\,°\text{F}$ ($-18\,°\text{C}$). I walked the eight blocks to school every morning wondering why anyone would want to live in a place like this. It was clear that I was now living in a different world altogether, a stark contrast to the hot summer days of the Egyptian Nile valley and the eternal sun baking the shores of the Red Sea. The warmth of my childhood home was far behind me, and a new, cold frontier was on the horizon. The bitter cold and the early sunsets sucked the energy out of my body, and as so many animals do, I wanted to hibernate—to sleep right through the darkness and icy winds ripping across the bare prairies, and awaken again only when spring brought back life.

As tempting as it sounds, humans are not built to hibernate. Remaining asleep for such a long period would have detrimental effects on our bodies: we must eat every day to replenish our bodies' supply of nutrition, and without regular use, our muscles begin to atrophy as they get digested away. But bears, groundhogs, skunks, and raccoons are well adapted to sleeping through winter. We still know little about this strange way of escaping the cold, but one thing is certain: proteins are crucial to regulating the massive metabolic transformation an animal must endure to adjust to a long slumber. Still, hibernation is costly. Bears, for example, can lose over a hundred pounds, roughly 30 to 40 percent of their body mass, during the winter. This weight loss is mostly the result of consuming fat reserves that the bear had built up during the summer and fall.

Remarkably, when a hibernating animal wakes up, its muscles are ready to work, unharmed by prolonged dormancy. During hibernation, a special protein ensures that muscles are not broken down. As the bear begins to prepare for sleep, alpha-2-macroglobulin is one of the many proteins released into the bloodstream.[5] This protein protects the bear's muscles from atrophy during hibernation by blocking the enzymes that normally chew up inactive muscle fibers. Meanwhile, other proteins, known as lipases, are directed to break down the fat reserves to supply the animal with the energy it needs to keep its heart pumping and its temperature from plummeting during the long sleep.[6]

The hiding spots of hibernating animals protect them and their young not just from the cold, but also from predators that might fancy an easy (and drowsy) prey. But in the Arctic Ocean, the open waters offer no such shelter or opportunity for hibernation. In the frigid waters of the Arctic lives a relative of the Moses sole known as the winter flounder. This flatfish thrives in the cold ocean, growing over two feet in length. Although the temperature is below freezing, most of the ocean water remains liquid—the salt in the ocean allows it to remain liquid at several degrees below the normal freezing point. Whales and other cold-water sea mammals carry an insulating layer of fat to shield their bodies from cold ocean water, but the winter flounder does not have this option. Nor does it have the high salt concentration that prevents the ocean from freezing below 0 °C. This means the fish is in grave danger from ice crystals. Ice crystals are unforgiving. They are unstoppable once they begin to form. They start as small "seeds" that then come together and grow into long dagger-like structures, piercing

Bears use a variety of proteins to regulate their metabolism during hibernation. One of the proteins that helps prevent muscle atrophy during their long sleep is alpha-2-macroglobulin (PDB code 1AYO), a small portion of which is shown here.

cells and damaging tissue. This is what causes frostbite. Yet the winter flounder's cells are unharmed in freezing water due to a special set of molecules known as antifreeze proteins.[7] Many arctic fish species, including sea raven, herring, smelt, and ocean pout, also have antifreeze proteins, which allow them to live in the frigid environment.

Even the coldest ocean water is only about two degrees below freezing. But on land, temperatures are often much lower. Some insects, such as the mealworm beetle and the snow flea, have developed their own versions of antifreeze proteins.[8] These proteins must work even harder to prevent frost damage. The mealworm antifreeze protein is over a hundred times more active than fish antifreeze proteins and can keep ice crystals from growing at six degrees below freezing. Even more impressive, the longhorn beetle antifreeze protein, along with other adaptations, allows this insect to thrive in temperatures as low as −25 °C (−13 °F). If we zoom out from the individual molecules of proteins and water and consider an entire species, we see that the ability to make antifreeze proteins provides an exceptional survival advantage. Such proteins can allow an organism to expand its geographical distribution to include environments that are hostile to many other organisms, that is, where there is less competition for food and less threat of predation.

Antifreeze proteins work by taking advantage of subtle differences in structure between liquid water and ice. Water is one of the most fascinating substances on Earth and indeed has some of the most intriguing properties of any liquid we know of. Common wisdom tells us that when things are heated they expand, and when cooled they shrink, and in fact, most liquid molecules draw closer as they cool. But when water freezes, it expands as the molecules move farther apart, forming

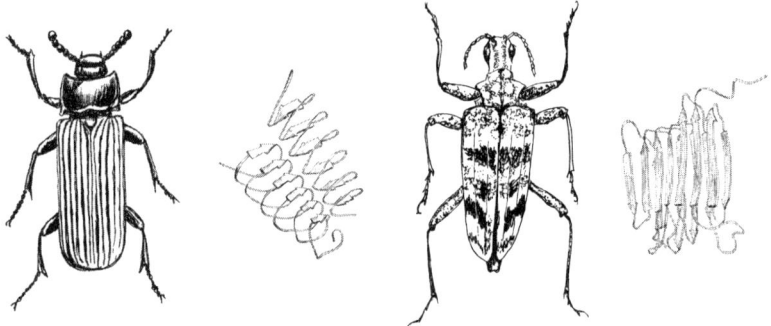

The mealworm beetle produces an antifreeze protein that allows it to survive in freezing environments (PDB code 1EZG). The longhorn beetle also produces an antifreeze protein (PDB code 6XNR) that, along with other adaptations, allows it to live in temperatures near −25 °F (−13 °C).

a lattice of repeating hexagons. This slight change in the way the individual water molecules interact with each other is why ice cubes float in a glass of water; this is also how antifreeze proteins protect fish from tissue damage. Antifreeze proteins can spot the subtle changes in the structure and arrangement of water molecules as they transition from liquid to solid ice. They seek out the tiniest ice crystal seeds within cells and surround them as soon as they begin to form, keeping the crystals from joining into larger, deadly icicles. In essence, antifreeze proteins act like a protective fence or a barrier that separates the tiny ice crystals from the rest of the liquid water in the animal's cell.

Fish and insects are not the only organisms with antifreeze proteins. Over sixty different plants, from large trees such as maples to smaller plants such as the perennial ryegrass, rely on antifreeze proteins to survive the cold.[9] This includes many familiar vegetables. Carrots, for instance, can remain in the ground for two consecutive seasons. Antifreeze proteins prevent them from freezing solid even as the ground around them ices up during the winter. When spring arrives, the carrots are ready to send up leaves and grow for another season. Antifreeze proteins have also been found in tiny ocean algae known as diatoms, in snow mold fungus, and in shiitake mushrooms. Even some microorganisms have antifreeze proteins. During a Chilean Antarctic expedition in 2016, scientists collected ice samples containing several species of bacteria to learn how microbes could survive in such extreme conditions. After repeated freezing and thawing of the samples, the bacterial species that survived were found to have their own versions of antifreeze proteins, which help them thrive in

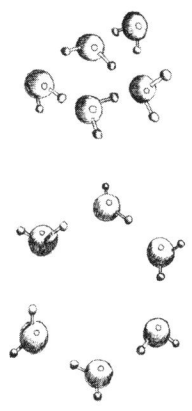

Water in liquid form *(top)* can take on different arrangements as the water molecules move around facing each other in different ways. In ice form *(bottom)*, water comes together to form arranged hexagonal structures when it freezes. Antifreeze proteins recognize the differences between the two forms of water and preferentially bind to the ice form to stop the spread of ice.

some of the most inhospitable environments on the planet. Bacteria and other unicellular organisms produce and secrete these unique proteins to provide a layer of protection from ice crystals.[10]

Each variation of the antifreeze protein is made possible by differences in structure. For instance, the winter flounder antifreeze protein is a single long helix that resembles a spiral staircase. It forms a barrier much like a speed bump that prevents the ice sheets from growing along a specific plane. By contrast, the longhorn beetle antifreeze protein is composed of two layers of switchbacks made from one continuous chain known as a beta helix.[11] This gives the protein a highly ordered structure, which complements the repeating hexagonal rings of the ice crystal, preventing the ice from expanding. The perennial ryegrass antifreeze protein has yet another arrangement composed of a single chain that coils into a wiggly spiral known as a beta roll. Whereas the single helix of the winter flounder is tight and narrow, twisting and climbing like a spiral staircase, the ryegrass protein makes much wider turns and twists in the opposite direction. Antifreeze proteins found in other organisms take on very different shapes, yet they all accomplish the same task.[12] Nature has found many ways to solve the same problem and provide a survival advantage to a wide variety of organisms. This is an example of what's known as "convergent evolution."

HOW TO SURVIVE BEING FROZEN SOLID

Because our fragile human cells lack the ability to make antifreeze proteins, they are no match for the cold, eventually succumbing to the painful damage of frostbite. But we are not completely helpless; through human ingenuity, we developed our own ways to fight the cold. Long before central heating, our ancestors ventured north into the coldest parts of Eurasia thanks to the invention of clothes made from animal hides. We also perfected our most unique skill of all—mastering fire—so that we could venture deep into the otherwise inhospitable bitter cold.

Without the ability to build a fire or invent central heating, organisms across the world have found other creative ways to deal with the cold. We've already explained some of them: sleeping the winters away, putting on thick layers of fat, or using antifreeze proteins to protect the cells from ice-crystal damage. Yet some organisms choose a far more unusual strategy to survive cold temperatures. In the north woods of Canada, as autumn passes and the coldest winter months approach, the wood frog will leave its pond to hide under a pile of dead leaves. This will be its home throughout the cold northern winter. Unlike the flounder or beetles, wood frogs do not try to keep their bloodstream from freezing; instead, they try to survive being frozen solid. When temperatures drop below freezing, wood frogs freeze completely from the inside out, turning into a solid chunk, like something you pulled out of your freezer.

The subfreezing temperatures preserve the frogs in a cryogenic state. But what about those razor-sharp ice crystals, which can slice open cells and irreversibly damage organs? The frogs protect themselves from this damage with the help of a set of proteins known as "ice nucleating proteins." Unlike the antifreeze proteins of the flounder and the beetles, which stop the growth of ice, the frog's ice nucleating proteins actually promote the formation of ice.[13] But instead of allowing the ice to grow in all directions, ice nucleating proteins direct the growth of ice crystals toward the outside of the frog's cells, effectively encasing each cell in an icy crystal ball that preserves its contents throughout the long arctic winter.

Still, even molecules as amazing as proteins can't protect the frog from freezing without the help of another essential molecule: glucose. This sugar is typically a source of food for most organisms, including humans. For a frozen wood frog, glucose works together with ice nucleating proteins to protect the frogs from frostbite. Come winter, the frogs pack their cells with massive amounts of glucose. The high concentration of glucose gives the inside of the cells the consistency of maple syrup and dramatically lowers the freezing temperature of the water that is also there. As the nucleating proteins outside the cells encase them in an icy shell, the large amounts of glucose inside the cells protect their contents from damage. With this combination of proteins and glucose, the entire frog freezes solid, yet its cells remain alive, awaiting the return of warmer weather. With a spring thaw, the entire process happens in reverse, and the frog is ready to bounce and croak, jump into ponds of melted snow, and find mates.

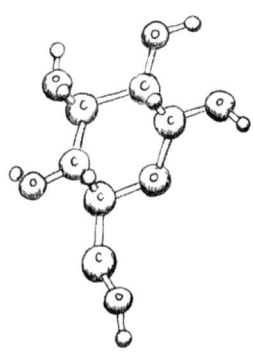

For most animals and plants, glucose is used as a food source. But the wood frog uses it to survive subfreezing temperatures. A high concentration of glucose inside the frog's cells keeps the insides from freezing while each cell is surrounded by a protective layer of ice.

In recent years, molecular biologists have deciphered the structures of many antifreeze proteins, showing how they keep a wide variety of organisms alive in some of the most inhospitable environments. But antifreeze proteins are not only fascinating for what they do for these organisms; they may also hold the key to solving some of the world's most pressing challenges. With their ability to prevent tissue damage by blocking ice formation, antifreeze proteins are now being investigated for their potential use as powerful cryoprotective agents.[14] Antifreeze proteins may be used in the future to prevent the wilting of fruits and vegetables that results from long-term freezing. By introducing antifreeze proteins into sensitive plants, we may be able to enhance their resilience to frost, ensuring more reliable harvests in regions susceptible to cold weather. Moreover, by maintaining the freshness of food, antifreeze proteins are poised to play a major role in reducing food waste during storage and transportation, potentially saving billions of dollars, increasing food security, and providing more reliable food supplies. Just as important, these proteins might also hold the key to extending the shelf life of donor organs, offering hope to countless individuals waiting for life-saving transplants.

As we look through the countless number of proteins made by all the different species in the world, we continue to find new uses for them. Just as antifreeze proteins can be repurposed to protect tissue from damage and crops from frost, other proteins have found their ways into myriad biomedical applications. For example, pardaxin, the shark repellent made by the Moses sole, has been found to have antimicrobial properties. The tiny protein, it seems, can help wounds heal better and can even fight flesh-eating bacterial infections—leading this small protein from an obscure, strange-looking fish from the Red Sea to become a promising new treatment for managing some of the most severe infections. Pardaxin may also have cancer-fighting properties.[15] Recently, when scientists added pardaxin to a mixture of healthy cells and cancer cells, the fish protein selectively targeted the cancer cells for destruction while leaving the healthy cells unharmed.

Pardaxin and antifreeze proteins are only two of countless examples of proteins that are being repurposed for a wide range of applications. Scientists are now beginning to reengineer proteins, reconstructing their basic building blocks to produce new proteins the world has never seen. These novel proteins are providing new avenues for treating infectious diseases and targeting tumors for destruction. Some are being used to seek and destroy environmental pollutants, capture carbon dioxide from the atmosphere, and even recycle plastic waste. These new additions to nature's vast repertoire of proteins have already proven to be powerful new tools for helping solve some of the world's most important, and pressing, problems.

2

BIRTH
MAGGIE M. FINK

THE BLACK WIDOW SPIDER, with the characteristic red violin on her belly, has always terrified me. As one of only two poisonous spiders in Northern Indiana, where I live, she is infamous. Even beyond the Midwest, her black bulbous belly and spindly legs are synonymous with the otherworldliness of spiders that causes so many to live in fear of them. For years, I imagined that any harmless eight-legged creature I found lurking in corners or emerging out of the grass was this ominous, poisonous spider. I refused to acknowledge that spiders could be helpful cohabitors and did not hesitate to find the nearest shoe and smash them dead.

Then I befriended a spider. An orb weaver, the largest I had ever seen. She made her home in the window of my home, a house surrounded by cornfields and mulberry trees. I met her only a few days after I brought my newborn son home from the hospital. She began to greet me every evening, just as the sun settled behind the farmhouse across the street. I was frightened at first by this unwelcome creature suddenly setting up camp in the window between my baby's crib and my rocking chair. I wanted to kill her, but I couldn't bring myself to do so after that first night of watching her spin her web. She floated on invisible lines, making a trap for other insects to stumble into once the sun set. Still, in my delirious new-mom fog, I had nightmares about her skeleton legs sneaking inside the house and crawling across my newborn baby. I hadn't learned yet that she and I wanted the same thing. I would rush into his room only to find her perched in her microscopic corner of the world, faithfully watching over my son, immoveable except for the slight sway from a summer wind.

Soon my dazed dreams stopped, and we synced our rituals. Every night, as I settled into my chair to nurse my baby, she would appear. Methodically, she would eat her web from the night before, a source of nutrients she would quickly recycle into that night's work of art. Her home would soon take shape: first the wide bridge, across the top of the window, then the anchors all around the edge. Once the framework was secure, she added lines branching out from the center to just the right spots along the circumference. Then she began her first of two dizzying trots around the circle, adding a final thread of structural silk and finishing it off with her tightly wound latticework of sticky silk to ensure her prey could not escape. With her masterpiece finished, she waited in the center, motionless. Every night I watched her while I rocked my baby to sleep. Every night she settled in after her work, just as I settled into the quiet work of mothering. Throughout the night, as I woke up to a fussing infant wanting to be changed, fed, or held, my spider would always be there, at attention, in the center of her home. With the morning sun's arrival, she would climb up the center of her web and spend the day crouched beneath the rain gutters, resting up for another night of vigilance.

One evening, just as the fall weather was settling in and my baby was showing off his smiles and holding his head strong, she didn't return to her window. For days, her web sat empty and began to give in to the wind and fall leaves crashing in, the broken weaving hanging loose from the corners of the window. Missing her company, I went looking for her. That's when I noticed, just below the windowsill, an egg sac, safely wrapped up in her silk so that it would survive the cold Midwest winter.

The silk my spider produced from the spinnerets on her body could produce different fibers, durable enough to hold her home and protect her offspring. In fact, an orb weaver can make up to seven different types of silk, each with unique properties. Some are sticky for anchoring the web to the side of a house or catching prey; some are strong, like those that form the main scaffold of the web; and others are pliant, allowing for give as the wind blows through the web. Each type of silk is made of a unique protein that forms the building blocks of the long silk strands. This tiny protein, known as spidroin, is roughly ten thousand times smaller than the width of the period at the end of this sentence. Yet when strung together, one after the other, the spidroin chain of proteins is one of most amazing biomaterials in the world. Spider silk is five times stronger than steel and almost three times tougher than Kevlar, while weighing far less than either of those human-made materials. It is also waterproof and can even conduct electricity.[1]

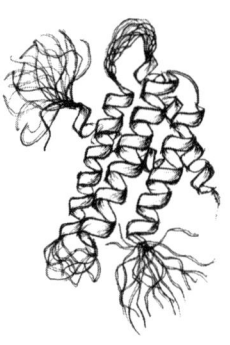

Spider silk is composed of spidroin proteins, a portion of which is depicted here. This section of the spidroin protein consists of five helical structures (PDB code 2LPJ). The individual proteins connect together to form the long strands of spider silk, which can vary in purpose, strength, and flexibility.

With countless spidroin proteins strung together in long fibers, the individual proteins fit like puzzle pieces to make an ordered structure. Thin fibers made of strings composed of spidroin align side to side to form thicker bundles, which the spider instinctively weaves together as she moves along the side of a house or other structure. Even more remarkable is the spider's ability to change the type of silk she produces on the spot to fit her immediate needs. Slight variations on the core structure of the spidroin give it different stickiness and strength, allowing the spider to build a home, catch prey, and protect her offspring. At the same time, the edges of the protein remain the same, allowing the proteins to fit together, end to end, as the chain grows.[2] By combining different forms of the spidroin protein as building blocks, the spider creates an ordered structure with predictable and reliable properties. This gives her an immense survival advantage.

The experiences we have as humans moving through the world are often mirrored in nature, even at the microscopic level. My spider friend was driven by the same instincts I am—and the need for structure, adaptability, and reliability in a constantly moving world is also true for the worlds inside our own cells. It is easy enough to observe the ways that nature has created its own ordered structures: think of a spider web or even a beehive. But order inside our own cells is much more elusive. Even the simplest yeast cells contain around 42 million protein molecules, each one busy doing its job to keep the cell healthy and functioning.[3] The number of molecules that must exist and events that must occur to keep us alive seems impossible, considering how each of the proteins needs to find the right place within the cell to perform its job. How does each cell not collapse into a massive traffic jam or tangled knot? The image of a cell in a biology textbook makes the cell's interior seem very simple. Uncrowded. Everything neatly packed into an unmoving oval. But in fact the cell is dynamic, meaning there is constant change and movement. An endless dance. Without order and stillness, there is no way these cells could survive, let alone allow us to exist in nature as complex dynamic beings.

CELL BONES

If you were to look at any single cell under a powerful microscope, much of the inside would be obscured by long fibers stretching from end to end, supporting an ever-changing environment. To the human eye, this web might not seem as organized as an orb weaver's, but actually it is far more complex: the fibers, structural proteins that resemble spider silk, are a stunning, intricately woven tapestry with an essential purpose. Because a cell is mostly water, structural proteins offer rigidity and stillness, providing the scaffolding needed for everything else in the cell to do its job. They make up the framework of the cell in a network known as the *cytoskeleton* (literally, "cell bones").

Much like a house needs wood framing for support, these strong, filamentous proteins support the fluid-filled cells of our bodies. But

structural proteins also work together to give each cell its shape and ensure it can divide, contract, transport cargo, and even trap invading
bacteria. An interplay of stillness and movement keeps each cell connected to the world around it. Cytoskeleton proteins extend through
the extremely long bodies of nerve cells, allowing some of them to reach
from the spinal cord all the way down to our extremities, a distance of
more than a meter.[4] In smaller cells that don't have the far-reaching effects of nerve cells, the cytoskeleton simply extends from the nucleus
in the center of the cell to the edge of the cell surface.

The cell bones of the cytoskeleton come in different forms. Like
spider silk, all cytoskeleton proteins grow from tiny protein building
blocks that assemble in a specific pattern into long chains. These chains
provide scaffolds for cellular operations and give order to the millions
of proteins swimming within each cell. One of the main structures responsible for providing this necessary order is made of proteins called
intermediate filaments, named for their size compared to other cytoskeleton structures. The individual proteins that make up the intermediate filaments have a simple helix-shaped rod, much like a spring.[5] As
they assemble, the springs begin to tangle together to form a ropelike
structure, coiling together into long threads. This filament grows larger
and larger as more coils begin to intertwine. Eventually they form a
bundle of rods that creates its final width of around six to twelve nano-
meters, about the thickness of a single strand of DNA.[6]

Assembly of these cell bones is easy to imagine. Much like braiding
hair or intertwining sewing thread, the effect of adding threads or
strands is to make the entire fiber stronger, a necessary feature of a
structure that provides support. For these intermediate filaments, chemistry drives the coiling, and the coils that result are decorated with
various atoms, like oxygens and carbons. Depending on the organization
of these atoms throughout the protein, some regions of the coil prefer
to interact with the water-filled cell, while other regions pair up with
the atoms of a nearby coil through various chemical attractions. In the
chaotic cell environment, this chemistry is enough to generate the order
necessary for intermediate filament formation.

The similarities between thread and intermediate filaments begin to break down, however, when we examine how these proteins manage physical stress inside and outside of the cell. If you pull hard enough on either end of a thread in say, your clothing, it will simply snap in half. Threads, by design, have little elasticity because their job is to keep the seams of garments in place, ensuring that, say, your shirt keeps its shape as a shirt. But as a cell moves and squirms about, intermediate filaments must be able to both keep the integrity of the cell's shape (like a thread) and stretch without breaking (like a rubber band, which doesn't help anything keep its shape). Somehow the proteins that make up intermediate filaments must have remarkable strength as well as the ability to be stretched and pulled without breaking.

The secret to the unique properties of intermediate filaments eluded scientists for years. After all, testing the physics of molecules as small as proteins can be a tedious endeavor because of the difficulty of measuring and visualizing such phenomena on a microscopic scale. But with the introduction of computational biology and other new technologies over the past several decades, scientists have been able to predict and test how exactly intermediate filaments perform the task of being both rigid and flexible. As a cell begins to migrate and change shape, great stress begins to act on the cytoskeleton. The interwoven network of intermediate filaments stretches and lengthens thanks to its coiled structure.[7] The strain of a dancing cell pulls on the spring, stretching it to its max. But just when you think it will break, the coil transforms into an entirely new shape. The individual proteins turn into sheets— flat stretches of chemicals that expand each filament to approximately 300 percent of its original length.[8] This transformation, which occurs by rearranging the chemistry along the proteins, also increases the stability of the overall structure. The balance that the proteins achieve between flexibility (of the coils) and strength (of the flat sheets) gives cells the ability to adapt to an always changing and sometimes stressful environment.

Intermediate filaments are made up of different individual proteins whose chemical and physical properties vary depending on the cell in which they are located. Their overall shapes, however, are strikingly similar. Vimentin is the most common protein used to assemble an intermediate filament. It can be found in a variety of animal cells and even in some bacteria. In humans, vimentin provides support to organelles like mitochondria or the nucleus, where DNA lives, anchoring them in the turbid fluid ocean of the cells. These filaments are also ubiquitous in developing embryos, where they assist in the transformation of a newly fertilized egg into specialized cells that come together to form tissues and organs.[9]

Keratins are perhaps the most recognizable of the intermediate filament proteins. They are essential parts of our hair and nails, which is why if you walk down an aisle filled with hair products, you will see "keratin" printed across the front of many bottles. Unlike other intermediate filaments in the cytoskeleton, which maintain great elasticity, keratin proteins need to be able to handle much more strain. When my baby's hands would wrap around my hair and tug relentlessly during our nightly nursing routine, the strands wouldn't break. Instead, they would resist his grasp, keeping my sweet boy and me connected. Much like the tight grip of my baby's hands on my hair, the bonds that link keratin proteins to one another are much stronger than those found in other intermediate filaments. Instead of the coils being decorated with mostly carbons and oxygens, as is the case in other intermediate filaments, keratins have many sulfurs.[10] Not only is sulfur the reason for the characteristic smell of burning hair; these sulfurs also link together to form

Keratin filaments are elongated, coiled strands. Here, two keratin coils form a coiled-coil structure (PDB code 6ECo). Interactions like these between the two coils help them form a robust network that contributes to the stability and resilience of keratin fibers found in hair, nails, and scales.

much stronger chemical attachments than those between the hydrogens found in carbon and oxygen.

Keratin filaments can undergo the same shape-shifting as other intermediate filaments: when they are stretched, their tightly woven coils turn into flat sheets. But the binding together of sulfurs helps the network stay intact under much greater mechanical stress. These sulfurs increase in number as the demand for stability increases, which accounts for the difference between "soft" and "hard" keratins. In our skin, and the skin of all mammals, soft keratins fill the cells, providing a physical barrier. With few sulfurs, they exhibit more elasticity. Hard keratins have more sulfurs, which provide stronger links, forming a rigid scaffold for hair as well as nails, horns, and scales.

THE TIGHTROPE WALKERS

Perhaps the most striking example of order inside our cells is found within the long muscle fibers responsible for moving our bodies. Muscle cells are unique in how they translate energy from the food we consume into orchestrated movements the moment we decide to walk, or run, or reach for something on a shelf. Skeletal muscle cells are very long, stretching to more than an inch in length.[11] Yet they are also very skinny, roughly a thousand times as long as they are wide. Together, thousands of long muscle cells form strong fibers that enable us to move, bend at the joints, and run away from danger. The highly coordinated movement of muscles relies on the ordered structures within the muscle cells themselves, an order that comes from the precise arrangement of proteins. The proteins stretch along the skinny body of the muscle cell in long filaments that, much like spider silk, are made of individual protein building blocks strung together end to end.

The main filaments in muscle cells are composed mostly of two proteins: actin and myosin.[12] The long filaments of actin and myosin sit side by side along the length of the muscle cell. Individual actin building blocks are tiny, each about the size of a spidroin protein. And like the spidroin protein, actins form a long, twisted spiral "rope" that can stretch to several millimeters in length. Myosin is somewhat thicker than actin and forms denser woven bundles that stretch alongside the actin filaments. Together, actin filaments and myosin bundles act as two ropes, sitting side by side within a muscle cell. Their movement relative to each other is the key to muscle movement. As the myosin bundles slide along the actin filament, the two far ends of the cell are brought closer, shrinking the length of the muscle cell and causing the muscle to contract. When the two ropes let go of each other, they slide back into their original places and the muscle relaxes, returning to its original length.[13]

How the two protein ropes slide along each other depends on the interaction between the actin and myosin, which in turn is a direct result of the unique shape of the myosin protein. Myosins are built to walk along the actin filament. They have a long, stringy body, which they weave together with other myosins to form the strong myosin bundle.[14] But they also have an appendage that resembles a human "foot." Within a myosin bundle, thousands of "feet" stick out, each looking for actin ropes to walk along. The feet of the myosin bundle "walk" along the actin fibers in a highly orchestrated march, moving

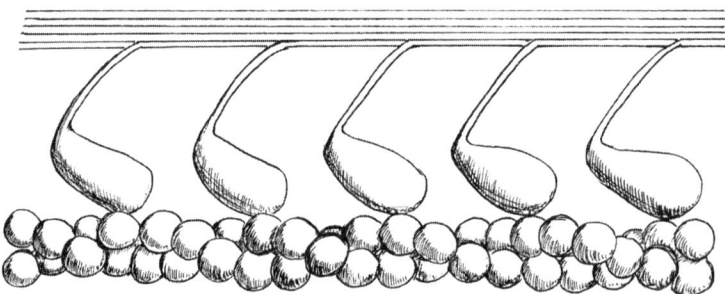

Myosin motor proteins can "walk" along long strands of actin, enabling cellular movement and muscle contraction. The actin filaments are depicted here as long, parallel strands forming thick bundles, while the myosin proteins are shown with their characteristic head and tail regions.

their way along the actin rope in a fraction of a second and making the muscle cell shorter every time a muscle contracts. This process requires energy in the form of adenosine triphosphate (ATP), a key molecule that acts as the universal energy currency of all living cells. In our bodies, energy from the food we consume is converted into ATP to supply the muscles with the energy necessary for myosin to walk along actin each time we move. Because of its ability to use energy to move in a specific direction, myosin belongs to a class of proteins known as motor proteins.[15]

While actin and myosin in muscle cells perform a highly specialized job when it comes to muscle contraction, the two proteins are not unique to muscle cells. In fact, actin and myosin are found in every one of our cells; but in other cell types, actin and myosin serve different purposes. Using the same tightrope-walking strategy, individual myosin proteins can slide along actin filaments to transport cargo—other proteins, fats, and even massive complexes that contain cellular machinery and RNA—from one side of the cell to another. In this case, the long stringy end of myosin is not tied up in thick bundles like those found in muscle cells. Instead, it acts like hands, grabbing onto cellular cargo while the "foot" end walks along the actin filament. And myosins are not only expert tightrope walkers; they are also heavyweight lifters. In skin cells, for example, a myosin protein can carry sacks full of the skin pigment melanin. Each melanin-filled sack (called a vesicle) can be thousands of times larger than a single myosin protein. This would be equivalent to a human walking a tightrope while carrying a blue whale on their back.

At the heart of this intricate machinery lies the globular shape of actin and its ability to form filamentous tracks for molecular cargo transport. The 3D architecture of actin filaments creates a dynamic road network within the cell, guiding motor proteins like myosin along its helical pathways. As cargo-carrying myosin motors traverse this intricate filamentous landscape, their finely tuned structure facilitates the attachment and detachment cycles of their "feet" that are required for efficient molecular transport. This dance between actin and myosin, orchestrated by their complementary structures, propels cellular cargo to its assigned destinations.

If we envision each cell not as a part that makes up the whole but as an entity unto itself, we can see why it needs multifunctional components. As complex beings, we have the luxury of our bones functioning only as bones. But the skeleton of a cell must do much more than just maintain a shape. The ability of the cytoskeleton to carry out all these functions is possible because proteins can shift quickly and seamlessly between assembly and disassembly in the fast-paced, crowded environment of the cell. Additionally, the expansive nature of cytoskeleton proteins across the cell enables individual protein components to interact with different regions of the cytoskeleton. Decorated with a host of proteins and molecules, the cytoskeleton can localize its changes within the cell in response to environmental stimuli. By maintaining the shape of the cell and acting as one of the "senses" inside, the proteins that make up the cytoskeleton not only save energy by limiting resources only to where they are needed but also help keep things as neat and orderly as possible in a cramped and chaotic cell.

FROM ONE TO TWO

The cytoskeleton has a third component in addition to actin and intermediate filaments: microtubules, slender yet robust tubes woven from protein threads. Scientists first visualized them in the early 1960s with an electron microscope, which uses electrons instead of light to capture images of things so small a normal microscope can't see them. (When electrons are accelerated to incredibly high speeds, they have wavelengths much shorter than those of visible light, enabling the microscope to achieve astonishing levels of resolution.) Yet when microtubules were first noticed, it was unclear what these structures were and what they were made of. It took a few more years of investigation into the role of microtubules for the individual components of these large filaments to be identified—and even then, the discovery happened by accident.

Only a year after microtubules were visualized, a graduate student at the University of Chicago named Gary Borisy was studying cell division. To do this, he used a drug called colchicine, which interferes with cell division and so allowed him to pause the division at various stages and investigate what molecular processes were affected. He wanted to know what colchicine was interacting with in the cell to have such a dramatic impact on cell division. The answer to this question would hopefully provide insight into what is required for splitting one cell into two.[16]

This question formed the basis of his thesis work. Borisy spent years trying to answer it, so long that even his adviser questioned whether Borisy's curiosity was leading him down a dead end. But Borisy persisted. Though the research was tedious, messy, and, at times, confusing, he was able to identify the target of colchicine. By breaking open cells and extracting everything the colchicine was sticking to, he found that the drug was binding to a protein, the same protein that formed the threads observed a year earlier by electron microscopy. Not only did Borisy discover what those strange strings were made of, but he also was able to show one of their primary purposes—pulling cells apart during their division.

Like intermediate filaments and other cytoskeleton structures, microtubules are constructed from individual proteins. Whereas actin is the smallest of the cytoskeleton structures, microtubules are the largest. Microtubules, as the name suggests, are tiny tubes that span the length and width of the cell. The building blocks of microtubules are a pair of proteins called alpha and beta tubulin. These proteins have an intrinsic attraction to one another—chemical forces bind them together into one unit. The unit begins to attract other pairs of alpha and beta tubulin subunits, which join end to end to form a chain. As the chain grows, it begins to spiral, forming a tube structure made of tens of thousands of individual subunits.[17] In many ways, microtubules serve a similar role as actin. They are superhighways that transport cargo across the vast ocean within a cell, with actin facilitating short-range transport and microtubules enabling movement across longer distances. They also help give each cell its distinctive shape. As the largest of the cytoskeleton structures, microtubules are the strongest of all the cell bones. They are also dynamic in nature—they can expand or contract as needed by adding or removing pairs of alpha and beta tubulin from one end or the other.

Dynamic instability is at the core of microtubules' adaptability, enabling them to grow and shrink in a controlled manner, much like the ebb and flow of a tide on a sandy shore. Instability initially seems like a disadvantage in biological systems. It suggests something fragile and prone to immediate dysfunction. But this instability gives tubulins the remarkable ability to switch roles between construction and deconstruction. This constant flux of growth and shrinkage not only grants microtubules their adaptability but also enables them to explore the cellular landscape with remarkable agility. Microtubules reach out like microscopic fingers to probe and connect with other structures, all while maintaining the delicate balance essential for the proper functioning of life's microscopic machinery.[18] Dynamic instability in microtubules is not just a peculiarity; it drives cellular life.

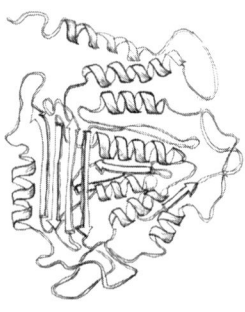

Microtubules are composed of two individual proteins, alpha and beta tubulin, which are nearly identical. Alpha tubulin (*top*) and beta tubulin bind forming a subunit (PDB code 3J7I), which can then begin to assemble with other subunits to form a microtubule (*bottom*).

Being dynamic is a necessary feature in the crowded and ever-changing cellular world. Individual molecular parts, like tubulin an-d actin, assemble to form the bones of the cell. But they also work together seamlessly and silently, allowing all the other machinery inside the cell to do their magnificent jobs. What looks like a chaotic intersection of hundreds of train tracks sure to cause a crash is instead an intricately coordinated scaffold and transport system that can adapt in real time to rapidly changing needs within the cell. These molecular railways also function as legs, with protein-assembled feet at the ends, that can move and relocate.

But just as quickly as these filaments assemble, they can be broken down—unzipped—to be put back together elsewhere in the cell as the ever-changing cellular environment makes new demands.

Cytoskeleton proteins may have different sizes, arrangements, or basic building blocks, but they all communicate in the language of chemistry, which enables the entire cell to function as one unit. The movements, assembly, and localization of microtubules, intermediate filaments, and actin must be coordinated at any moment to avoid the collapse of the complex biochemical dance within each of our cells. And at no time is this coordination more apparent than during the process of cell division.

The birth of a new cell requires that all three classes of cytoskeleton proteins cooperate. When a cell decides it is time to divide, a nearly perfect orchestration of players must come together to pull it off—or rather, pull it apart. This is when the cytoskeleton proteins temporarily set aside their day jobs of providing structure and transporting cargo to join hundreds of other proteins in the task of turning one cell into two. During cell replication, DNA must be duplicated in the mother cell and separated into two daughter cells. Normally, the DNA is housed inside the nucleus, which is anchored at the center of the cell by intermediate filaments. As the DNA is duplicated, the intermediate filaments holding the nucleus in place begin to disassemble, and the membrane separating the nucleus from the rest of the cell dissolves. With no barrier between the DNA and the rest of the cell, microtubules begin to grow, reaching toward the newly made DNA. The tubular structures start at opposite ends of the cell and find their way to the center of the X-shaped chromosomes, the large structures containing all our cells' DNA. As the strongest of all cytoskeleton structures, the microtubules tug at the center of each chromosome, tearing apart the two copies of the DNA. With one coordinated tug, the microtubule ropes drag the split chromosomes to opposite sides of the cell. Once this separation has taken place, actin filaments begin to form a ring around the center of the cell. As the ring shrinks, the cell is squeezed into two blobs, which eventually split into two separate cells. Once the process is complete, the DNA is rehoused in each cell's nucleus, again anchored by intermediate fila-ments. The microtubules and the actin recover from their cell-splitting roles and return to their usual jobs of giving cells their shape and trans-porting cargo.

It is easy to think of cell division as the creation of new material. After all, DNA is doubled, and one cell becomes two. But in many ways, it is simply a reorganization—a repurposing of what the cell already has. This is most evident in the versatility of cytoskeleton proteins. With their dynamic nature, they shapeshift as they transition from one job site to another. The arrival of a new cell is nothing more than a job re-assignment for proteins, which are no strangers to adaptability. Actin and microtubules moonlight as midwives for the birth of a daughter cell, then go back to being the cell bones and the superhighways for molec-ular cargo. Nature, as it were, has brought about versatile building blocks that can be rearranged for seemingly very different purposes.

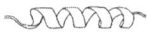

The house I was renting when my second baby was born was surrounded by trees and fields of corn. Across the street was a farmer who would wave at me every evening when I would walk around the front yard, bouncing my baby and exploring with my toddler. We watched the sea-sons change every year, marking time by the height of the corn, the sound of the tractor harvesting in the fall, and the arrival of snow. But the beginning of summer was my favorite time. In the back corner of my yard, at the edge of the cornfield, stood a mulberry tree. One morning, I ventured back there alone and found the ground bleeding with berries, dark purples and reds. I picked a few and ate them right off the branch. Later that day, with a baby strapped to my body and a toddler at my side, I visited the tree again. By the end of the afternoon, our hands and lips were stained purple, and a mulberry pie was cooling on the counter.

Halfway through the summer I noticed thin white threads ap-pearing around the branches, encasing what were now dead, brown leaves. When I looked closer, I saw a hoard of white silkworms, feasting on the mulberry leaves. Each time a breeze came through, they would wiggle and dance, then return to their banquet. After a summer rain, the silken threads would glisten in the sun. While these insects don't make silk as tough as a spider's, their silk is just as remarkable. With a single silk strand, up to three thousand feet long, the larvae make co-coons that protect the silkworm as it undergoes its metamorphosis into a moth.

Like spider silk, each fiber from the silkworm is made up of several different types of proteins that assemble to form the thread in a special gland. But as the silk proteins come together, they exhibit a more reg-ular, repetitive structure compared to their spider silk counterpart. This structural uniformity contributes to the silk's lustrous appearance and smooth texture, characteristics prized by artisans and textile enthusi-asts worldwide. While silkworm silk may lack the sheer strength of spider silk, its threads still possess remarkable tensile properties, making them perfect for weaving luxurious fabrics and garments. In recent years, too, scientists have discovered that these properties make silk-worm silk an ideal material for biomedical applications. In addition to being strong and smooth, silk proteins are highly biocompatible and bio-degradable, which make them well suited for medical uses. Because silk comes from a living organism as opposed to synthetic material, a relationship between silk and human tissues can form when they are used as bandages or sutures, significantly lowering the risk of the body reacting negatively. The silkworm silk can also serve as a versatile scaf-fold for engineering tissues ranging from bone and cartilage to skin and nerves. The porous structure and mechanical resilience of silk fibroin facilitate many cellular functions, mimicking the extracellular matrix environment essential for tissue development and repair.

As scientific exploration of silkworm silk's biomedical applications continues to unfold, the potential for using spider silk in regenerative medicine, drug delivery, and surgical materials has expanded as well. The diverse types of silk produced by spider species reflect the adapt-ability of spidroin proteins to various functions, an ideal trait.[19] Drag-line silk, used for the primary framework of a spider's web, boasts a combination of strength and elasticity. Capture silk, employed for en-snaring prey, is stickier and more viscous. Understanding the nuances of spider-silk protein structures has inspired scientists to try adapting these proteins, as well as novel biomaterials based on their unique properties, for use in new contexts. Researchers are exploring the development of artificial ligaments, and tendons, made from spider silk proteins. The lightweight yet robust nature of spider silk makes it an attractive candidate for creating durable textiles. And researchers are looking into the use of spider silk proteins in space exploration, where lightweight and strong materials are essential. Potential applications beyond planet Earth include the development of textiles and biomaterials that can withstand the harsh conditions of extraterrestrial environments.

Just as humans have repurposed the silk from spiders and worms, organisms have, throughout evolutionary history, taken a protein that carries beneficial properties and transformed it for an entirely new function. For instance, in the springtime, the spidroin filaments are the bones of the web and the superhighways for the spider. But come fall, the same spidroin protein moonlights as a surrogate mother, protecting a precious cargo of eggs, bringing forth new life from old. The proteins are not reinvented—they are simply repurposed in ways that save energy and other resources. It turns out that nature is extremely good at making use of what it has. Many of the proteins we find in complex multicellular organisms can be traced back to a protein with similar structures in bacteria, the ultimate recyclers.

THE SHARED DANCE OF LIFE

Hundreds of years before the development of the high-powered microscopes we use today, a Dutch cloth merchant was one of the first people ever to see a bacterium. Antonie van Leeuwenhoek was endlessly curious about the world around him, despite receiving no formal scientific training. He would spend his days examining the quality of linens in his shop with a handheld magnifying glass. A combination of business acumen and fascination with lenses led him to build his own microscopes in the 1660s. While others such as Leeuwenhoek's English contemporary, Robert Hooke, had already been developing their own microscopes, it became apparent that the merchant's lenses were far superior to anything that had been developed. A surviving microscope of his can magnify up to 275 times, and it is likely that others could magnify up to 500x.[20] Today's light microscopes can reach up to 1500x, but require more complex compound lenses, while Leeuwenhoek's only needed a single lens.

It was Leeuwenhoek's observations of the invisible world around him that gave him the title Father of Microbiology. One day, out of pure curiosity, Leeuwenhoek scraped a bit of the plaque off his teeth after not brushing them for three days straight. His account of what he found is delightful. In a letter to the Royal Society of London, he described seeing "many very little living animalcules, very prettily a-moving." He attributed their motility to thin little feet that protruded from their bodies. Fascinated and a bit terrified by the realization that there were living things all around him, Leeuwenhoek began to explore more samples, including his wife's dental plaque, pond water, crushed pepper, and even his own semen. Leeuwenhoek's window into the cellular world was an incredible scientific discovery for a man with no higher education or formal training; however, it would be another couple hundred years before Robert Koch's germ theory was accepted and the field of microbiology became rigorously studied. Even today our understanding of bacteria and their motility is evolving rapidly. But we know that the proteins that caused Leeuwenhoek's animalcules to dance under the microscope share many of the same properties and structures of those found in our own cytoskeleton.

The animalcules Leeuwenhoek saw were likely using flagella to swim under his lens. Flagella are long, whip-like structures jutting out of the bacteria that form essentially a nanomachine, a motor built with proteins. This engine allows many bacteria to swim and tumble, interacting with their environment to find food, friends, and foes. The motor itself is made up of several different types of proteins, called flagellin, that are anchored to the bacteria. Protruding from this engine is a protein appendage, a hollow twisted tube that mirrors the helical shape of the flagellin. Like intermediate filament proteins, the flagellin have similar helical shapes and chemical properties that allow them to self-assemble into this supramolecular appendage.[21] But the flagellum, which contains over thirty thousand flagellin proteins and is longer than the bacteria itself, must be assembled outside the cell.[22] To prevent a flagellum from forming inside the bacteria, likely causing it to burst, an individual flagellin protein does not adopt its final coiled shape inside the cell. Instead, the partially twisted flagellin is ejected through a tunnel in the bacterial cell membrane that is just the right size to accommodate it. Once the flagellin reaches the outside world, another protein comes along as a chaperone, making sure it is protected until it reaches its destination on the growing filament. This process continues until the full-length flagellum is completed and the bacterium can begin its run-and-tumble life.[23]

For the most part, flagella are one size fits all for the bacteria that have them. As far as researchers know, these nanomachines assemble and operate the same way across the bacterial kingdom. Even the flagellin protein itself has a nearly identical structure in all kinds of bacteria—four distinct coils, each called a domain, attached to a tangled knot of interwoven sheets and loops, the globular domain. The regions of the coils that are responsible for interacting with other flagellin proteins to self-assemble are also largely unchanged from species to species. Variability between species arises on the other coiled parts of the flagellin proteins that end up on the outside of the flagellum, but these proteins have no effect on how the flagellum functions mechanically. They enable bacteria to evolve and adapt to their environments by altering their chemistry. Importantly, they can sense food as well as competing organisms in the environment.[24] Such differences were propagated over and over throughout evolutionary history, linking movement with changes in the environment.

While bacterial flagella and the intermediate filaments found within the cells of multicellular organisms serve distinct purposes, their underlying molecular architecture exhibits striking parallels, underscoring the ingenious ways in which nature repurposes protein structures for diverse biological functions. Both structures share a common building block: the coiled-coil structure, or motif. This particular motif involves two or more coils wound around each other to form a stable structure resembling a twisted rope. In intermediate filaments, this coiled-coil arrangement contributes to the filament's strength and resilience, while in bacterial flagella, it forms the core structure of the flagellar filament, enabling its rotation and propulsion.

This raises an intriguing question: How did such structurally similar proteins find their way into seemingly unrelated cellular components? In the case of intermediate filaments and bacterial flagella, there is no evidence of a shared ancestry or common evolutionary precursor that might explain the striking similarities in their structural motifs. Still, the repurposing of protein structures exemplifies the economy of nature. Rather than creating an entire new structure, organisms can re-purpose the scaffolds they already have. As we will see, this ability to use existing molecular frameworks for diverse functions is true for many proteins involved in essential biological processes, and maybe even for those used in more complex experiences that make us distinctly human, namely, our consciousness.

CONDUCTING CONSCIOUSNESS

My baby, whom I had rocked and nursed next to the dancing spider, quickly moved through his chubby baby phase, wild toddler years, and first day of kindergarten. Now, a decade later, he is squarely in the incessant-questions stage of life, with his mind racing to learn more about the world all around us. Our weekly drive to piano lessons is usually when his questions come pouring out. What is the fastest animal? What is the name of that star, just above the moon? Which river is the oldest in the world? Is it okay to eat meat? What is pasta made from? Can I eat noodles if I'm a vegetarian? Won't eating plants hurt the bees? What happens if all the bees die? What happens when we die? Why are we here? How do we know we are real?

Answering his questions about the fastest animal or the oldest river is easy enough. Scientists have measured those phenomena. But the existential questions about consciousness and life are more difficult. These questions arise in the minds of nearly every human being. Religion and philosophy have tried to give meaning to them. But science has had a difficult time providing a biological answer to the question of consciousness, because science must be able to observe something to test it. Through neuroscience, we can explain how inside a brain, electrical currents fire and connect in brain cells, giving rise to thoughts, emotions, and memories. But a deeper question remains unanswered: How do we know that neurons and synapses are creating the emotion of fear, or that they are allowing me to recall the memory of my baby pressed against my chest night after night? Proving consciousness seems an impossible goal.

Some scientists, however, have proposed that proteins may play a critical role in consciousness. When Gary Borisy was conducting the experiments that would eventually identify tubulin as the primary component of microtubules, he found something interesting. As he was probing cells with colchicine, he noticed that brain cells contained large quantities of microtubules compared to other cells. At the time, Borisy believed that finding such an abundance of microtubules in brain cells was a mistake. Neurons don't divide, which meant that microtubules couldn't undergo the same rapid disassembly and assembly of the tubulin subunits. There would be no need for microtubules to be present in such large quantities in cells that didn't need to be pulled apart as they split into two. But neurons do have distinct shapes that allow them to send messages over long distances, and microtubules are essential to maintaining that structure. Their stability, in fact, may make them the perfect structures to support consciousness. The role of microtubules in consciousness is a central component of a controversial theory proposed by the physicist and anesthesiologist Roger Penrose. Renowned for his work on black holes and the nature of time, Penrose joined forces with Stuart Hameroff, an anesthesiologist with a fascination for altered states of consciousness, and throughout the 1990s they embarked on a quest to explore how microtubules, quantum physics, and neurobiology might explain our thoughts and awareness.

If we were to make microtubules "life size," their ability to generate consciousness would stop. How we move our arms and legs or bump into tables and doorways is vastly different from what happens when electrons bump into each other on proteins. The laws of physics don't operate the same way once we zoom in to the universe of proteins. The oxygen and hydrogen decorating the structures of tubulin contain electrons that behave by their own rules. Our ability to observe that behavior, or lack thereof, has led to a different type of physics called quantum mechanics. A whole new set of mathematics and physics dictates how these particles behave. This is most easily described by the paradox of Schrödinger's cat. In this thought experiment, a cat sealed in a box with a radioactively triggered poison or explosive device, all arranged so that the cat may or may not die inside, is considered to be both alive and dead. Once the box has been opened and the cat is observed, however, the cat is either alive or dead, but not both.

This experiment may seem strange, but it describes something important about quantum physics. In particular, if we replace the cat in a box with the electrons along a microtubule, we can start to understand how quantum mechanics can explain consciousness. Electrons exist in a state of quantum fuzziness. This fuzziness means that the more precisely we try to measure an electron's position, the less certain we become about its momentum, and vice versa. It's as if the universe itself imposes a fundamental limit on our ability to know everything about a particle simultaneously—or as if nature herself plays a game of hide-and-seek with us. As the electrons along the microtubule move around, they could be in any number of positions. But as they interact with each other along the microtubule, a quantum event occurs and the electrons are forced into an observable position. This collapse of the electron's quantum properties transfers information to the neurons that can then be fired throughout the nervous system.[25]

The resulting theory, called the Orchestrated Objective Reduction (Orch OR) theory, postulated that quantum events within microtubules play a crucial role in the emergence of consciousness. Penrose and Hameroff suggested that the orchestrated quantum vibrations within microtubules act as a sort of cosmic conductor, orchestrating the reduction of possibilities in the quantum realm and collapsing them into the singular reality we experience as conscious awareness. These quantum vibrations, the researchers proposed, synchronize with the activities of neurons, creating a harmonious interplay between the quantum and classical worlds. It's as if the brain, at its most fundamental level, operates like a quantum computer, processing information in ways that challenge our conventional understanding of biochemical processes. The microtubules, often thought of as mere structural components, take on a newfound significance in this theory, serving as the stage on which the complex questions of quantum consciousness unfold.

Penrose and Hameroff's theory of consciousness is difficult to wrap one's head around. Quantum mechanics itself is a complicated field, and entangling the already enigmatic idea of consciousness with biology and quantum physics is a lofty goal. For these reasons, proving the validity of this theory has not been easy. In fact, over the past several decades, experiments have disproven parts of the Orch OR theory, calling into question the role of microtubules in consciousness. Yet exploring ideas is part of the creative scientific process, even if those ideas turn out to be wrong. Many modern scientific theories that are currently accepted took decades upon decades to be proven true. Moreover, even if scientists never fully explain the fundamental questions of consciousness that have plagued many wise people and keep my ten-year-old up at night, one thing is clear: proteins are integral to who we are, who we perceive ourselves to be, and how we interact with the world around us.

3

AWAKENING

SHAHIR S. RIZK

IN THE SMALL WORKSHOP in Zagazig, Egypt, Mo was busy working a sewing machine. All around him, the walls were covered with colorful sheets of vinyl, nylon, leather, and suede. The latest shades of upholstery hung on the walls surrounding the nook where Mo sat. What little space was left on the walls was covered with posters of sports and luxury car models from the 1980s. Mo spent his days crafting fitted upholstery for car interiors in the shop he co-owned with his brother Sayeed. The two brothers' shop was the best in town, crafting perfectly fitted covers that matched the contours of each car seat. Sayeed ran the business side of the shop and had a small crew of workers who did minor car repairs. But the coveted seat covers were meticulously sewn by Mo, who could not see the colors of the fabric on the walls or even the sewing machine that was his work station. Mo had been blind since birth.

My father knew the brothers well and took his newly purchased pea-green 1977 Fiat model 131 to the brothers' shop for a custom set of seat covers. When we arrived, Mo immediately recognized my father's voice over the noise in the shop and left the sewing machine to greet him with a warm smile. In another corner, a couple of young technicians worked with Sayeed to replace a glass pane on a parked Buick. After catching up with my father, Mo walked over to the workers and quietly inspected the glass pane with his hands. He ran his palms over the slot in the door where the window would fit and declared that there was a problem. He was convinced that they had the wrong window for that car model, and that if they tried to install it, it would surely break. The young technicians exchanged smiles and eye rolls, quietly mocking Mo's claim. How could this blind man see such a thing? Mo walked back to his sewing machine a few yards away, and the young men continued with their work. They were able to install the glass in the door, but as soon as they rolled up the window, it cracked down the middle. Hearing the window break from his station, Mo let out a big laugh and yelled out, "I told you it would break!" As a ten-year-old, I watched in disbelief as the blind man showed everyone what they could not see. It was clear that Mo had a special talent. His hands were his eyes, and he could "see" better than those around him. What Mo had was a keen perception of the three-dimensional world that enabled him to craft the best fitted upholstery in town.

The ability to perceive the world and respond to it is one of the most fundamental properties of living things. Our eyes are the first and most important tools that we use for this job. The human eyes are truly incredible. Together, they collect roughly 80 percent of the information that our brains receive. Our eyes are extremely sensitive to light; in total darkness, they can detect a single photon, the smallest possible packet of light. It's no wonder that we rely heavily on visual cues to navigate the world. Light means clarity and safety, while darkness triggers feelings of unfamiliarity, mystery, and fear.

In the brothers' shop, my eyes were flooded with the light reflected from the colorful textiles hanging on the walls. From suede and vinyl to leather and nylon, each material reflected the buzzing neon lights in its own unique way, dazzling my ten-year-old eyes with sheens, tones, and shades that danced as the gentle summer breeze from the windows swayed the large sheets ever so lightly. Each light wave reflecting from the colorful materials entered my eye. As the light passed through the soft transparent tissue that makes up the lens at the front of my eye, muscles on the edges of my pupils moved in unison to stretch or compress the lens tissue, focusing the light beams as they entered. The beams of light then continued their journey through the liquid that fills my eyeballs and collided with the retina—the tissue that covers the back wall inside my eyes.

All along the surface of the retina, a group of so-called cone cells act as the collection stations for light, making them the work hubs for seeing color. The surface of each of the seven million or so cone cells is speckled with proteins known as opsins, which lie at the frontline of the complex process of vision.[1] By interacting directly with the light waves after they travel through the aperture of the eye, opsins act as tiny eyes within the eye itself. Opsins are miniscule—about 100 million times smaller than the cone cells on which they reside. In fact, if each of our eyes were to expand to the size of a large baseball stadium, each cone cell would be roughly the size of a hotdog bun, and each opsin would be much smaller than the poppy seeds on that bun. This means that an immense number of opsins must work together to collect enough light to make sense of our surroundings.

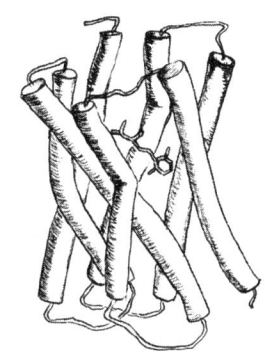

A vision receptor protein (PDB code 1F88) showing the characteristic basket structure. Inside the basket, a pigment changes shape when light is received, causing the protein to twist in response.

With a shape that resembles a basket, opsins collect light reflected from objects around us. But receiving information is not the only job they carry out. Opsins must also communicate this information to other proteins deep within the cone cells, which then relay it to neighboring cells, and so on, until the signal reaches the brain for processing. Whenever an opsin is activated by light, it twists and turns, contorting and bumping into surrounding proteins. This twisting leads to a cascade of molecular dances, much like a Rube Goldberg machine. At each step, the tiny signal from one protein is amplified as it travels to the next, creating a chain reaction. Each of the proteins in this chain relays the signal in its own way. Some proteins twist and change their shape; some drive chemical reactions; and others move charged ions of sodium, potassium, and calcium across membranes to create tiny voltages and electrical signals. In mere fractions of a second, these signals travel through the thick bundles of the optic nerve to our brain, where they are instantaneously translated into images of shapes and shadows, colors and tones.

Today, we know that opsins receive light reflected from objects in our field of vision. But for over a thousand years, it was believed that our eyes cast vision rays to scan our surroundings. Plato, Ptolemy, and many other Greek philosophers and mathematicians believed in this so-called emission theory. Euclid of Alexandria even devised sophisticated mathematical models for how rays emitted by our eyes illuminate objects we see. Anecdotal phenomena, like the eyes of a cat shining in the dark, seemed to support this misconception. But there was no rigorous testing of this idea until around 1000 CE, when the Arabic scientist Ḥasan Ibn al-Haytham conducted a series of seminal experiments revealing much of what we know about the field of optics.[2] Born in modern-day Iraq, Ibn al-Haytham moved to Cairo to pursue his research in physics, mathematics, and astronomy. Through systematic experiments, Ibn al-Haytham debunked the long-held emission theory and demonstrated that light is instead collected by the eyes and processed by the brain. His most important experiments came from an observation that he made while sitting in a dark room. Through a tiny hole in the wall, Ibn Al-Haytham noticed that the light from a lantern outside the room was projected on the opposite wall. Adding another lantern at a different height resulted in the projection of two images, each image coming from one of the lanterns. This led him to conclude that light was emitted by objects and received by the eyes and not vice versa, as the Greeks believed.

While this may seem like a simple experiment, at the time it was revolutionary. By positioning lanterns outside the dark room, Ibn Al-Haytham had constructed one of the first pinhole cameras. In his later studies of the human anatomy of vision, he proposed that our eyes work like pinhole cameras, receiving light through the tiny aperture of our pupils, and that images are projected on the back wall of the eyes.

To illustrate how this happens, he sketched the connection of the eyes to the optic nerves and brain in elaborate diagrams. He also documented the theory in a seven-volume book called *Kitāb Al-Manāẓir* (Book of optics). Later translated into Latin, the book influenced the ideas of Kepler and Galileo, leading to the invention of the telescope. Most important, however, by asking questions related to his observations and following them up with experiments, he helped lay the foundation for the scientific method as we know it.

In the twentieth century, scientists were able to zoom in on the molecular structures of opsins, revealing their characteristic basket structure. They also discovered the secret to the ability of opsins to collect light from the environment—a set of pigments embedded in the basket itself. These pigments are modified forms of the vitamin A we consume in foods like carrots and spinach. When light hits one of them, it causes a chemical reaction that changes the shape of the pigment from a bent structure to a straight formation. This shift in the structure of the pigment molecule, which is embedded deep inside the basket, twists the opsin protein, signaling that light has been received. Yet all the billions of opsins crowding the back of our eyes belong to only three color receptor types, corresponding to red, green, or blue. Each of the three opsins houses a unique pigment molecule, which reacts with one of these three colors. How, then, can the human eye see not just three colors, but an estimated ten million different colors?[3]

The key to color perception is not how each opsin works separately, but how all three work together. Instead of working as three independent channels for capturing light, our brain interprets the collective signals from the three opsins to perceive the precise shade of the color streaming through our pupils. As we observe the world, each shade of color activates the three opsins to a different degree. The combination of the signals from each opsin provides one collective signal that is a signature for a specific shade of color. For instance, looking at trees elicits a strong response from the opsin with the green pigment, and its light-induced dance is captured by a collection of proteins that surround the opsin. That movement is in turn relayed to the brain as the green leaves of the trees. A red apple, likewise, will trigger a twist in the red opsin, and red will be perceived. But a yellow banana will activate both the red and green opsins about equally, while a tangerine will activate the red opsin a little more than the green opsin, signaling orange. If any one of the opsins is not functioning properly, the balance in how colors are perceived will be disrupted, leading to color blindness. In fact, one of the most common causes of color blindness is the inability of the body to produce fully functional green or red opsins.[4] As a result, people with those types of color blindness have trouble distinguishing between shades of red and green.

Bees cannot see the color red, but they have opsins for seeing ultraviolet light. The flowers of some plants contain patterns that can be seen only with these ultraviolet receptors, in order to better attract bees. For example, the way we see a sunflower is more like the bottom of the drawing, while bees would see extra colors like the top of the drawing.

Even with three fully functional color opsins, we are still blind to so much. The range of colors we see amounts to only a tiny fraction of the entire light (electromagnetic) spectrum, which includes x-rays, ultraviolet light, and infrared light—none of which we can see. Moreover, not all animals can perceive colors in the same way. For instance, what we see as orange stripes on a tiger are to a gazelle indistinguishable from the tall green grasses of the savannah. By contrast, many birds and insects see ultraviolet light, which is just outside the range of our vision.[5] Bees cannot see red, but they can detect ultraviolet light reflected from flowers in their search for nectar. Crows, which to us look black and virtually indistinguishable from each other, appear to have unique and elaborate patterns of colors when viewed under ultraviolet light. Many birds use patterns like these to identify members of their flock and to find attractive mates.

Many organisms have more than three opsins. The record holder for having the most is the mantis shrimp, with twelve different opsins. The mantis shrimp, it seems, can see ultraviolet light and can even detect polarized light, light that vibrates at an angle. While we can distinguish colors only by how fast the light wave is vibrating, the proteins in the eye of the mantis shrimp can tell if the wave of a particular color vibrates up and down or left to right.[6] With so many more color opsins than we have, it is hard to imagine the variety of colors the mantis shrimp can see, though it is likely far more colors than our eyes can distinguish. But with many fewer brain cells than us, it remains a mystery how the mantis shrimp processes all the light information it receives from its surroundings.

It is not clear why the mantis shrimp needs to see such an immense number of colors and tones in the ocean, or why it can distinguish between different angles of polarized light. Seeing underwater, however, requires slightly different mechanisms than seeing on land. For example, light reflected from objects underwater tends to retain more of its polarization angle than light reflected in air does.[7] When presented with an object, the mantis shrimp's eyes rotate independently to align its receptors with the angle of the light waves bouncing off the object while simultaneously collecting information on its shape, color, and distance. Having many ways of seeing the world likely helps the mantis shrimp hunt for prey or detect danger. These extraordinary abilities also help the mantis shrimp navigate a world where camouflage is king. Many marine organisms, including the mantis shrimp, are experts at blending in with their environment. One of the most proficient is the octopus.

Without a protective shell or strong bones, an octopus may seem like easy prey. But octopuses have survived in the ocean for an estimated 330 million years. They pre-date modern sharks by 130 million years and dinosaurs by 80 million years. With a unique intelligence and the exquisite ability to blend into their surroundings, an octopus can both hide from predators and easily ambush prey, gaining a huge survival advantage in a hostile world. To help it accomplish this dazzling disappearing act, the skin of an octopus is packed with tiny balloons, called chromatophores, that are filled with colored dye molecules called chromophores. When chromatophores are squeezed by the muscles of the octopus, they stretch, spreading the dye molecules over a larger area and making the color appear lighter. Relaxing the muscles brings back the darker tones. A special protein in the cells of the octopus known as reflectin also helps the octopus change its color between shades of white and blue.[8] In response to external signals, groups of reflectin proteins stack together in different arrangements and scatter light to reveal shades of blue or bands of white. With the combined effort of proteins like reflectin and the dyes in the chromatophores, the octopus can match the color and texture of its environment within seconds.

Achieving a convincing camouflage requires the octopus to perceive the colors and shades in its environment. But as it turns out, the octopus's eyes are unable to see many of the colors it uses to form its disguise.[9] In fact, the octopus is mostly colorblind, able to perceive only a few colors through its eyes. Instead, opsins on the skin of the octopus receive light signals, mostly in shades of blue, and communicate to the chromophores how to stretch or contract to get the shade that matches the surroundings. It may seem strange that an organism can see with its skin, but the octopus's tentacles have billions of opsins that act as nanoscopic eyes, pointing in all directions at once. That's not to say that these opsins on the skin are "seeing" in the way we do. It is more likely that the opsins on the skin of the octopus "sense" light and react to it in a way that does not require the octopus to use its eyes to look in all different directions at once. In a way, the octopus uses its skin to sense its environment, much like Mo used his fingers to perceive the world and sew his elegant car seat covers.

LET THERE BE LIGHT

Light perception has been a characteristic of organisms since life arose on our planet. Ancient forms of light reception emerged long before animals and plants split from our unicellular predecessors. Tiny bacteria, and their more ancient cousins, archaea, have specialized opsins that detect and react to light. Perhaps one of the most striking examples is a protein from a peculiar microorganism that may have been around for nearly two billion years. Haloarchaea live in extremely salty environments, where very few other organisms can survive. Vast areas of salt lakes inhabited by haloarchaea, also known as halobacteria, look purple from space due to the purple pigment embedded in their photoreceptor

protein. These microbes can detect light using a protein known as bacteriorhodopsin, which is similar in structure to our own opsins.[10] But as
far as we know, microbes don't use this photoreceptor protein the way
we use ours—to produce a color picture of our surroundings. Instead,
they use it to capture light energy, much like a solar panel.

Like our own opsins, the light captured by the purple pigment
within bacteriorhodopsin triggers a photochemical reaction that
changes the shape of the pigment. As the pigment relaxes back into its
original structure, the energy is used to move positively charged protons across one of the haloarchaea's two membranes, creating a voltage.
When the protons accumulate on one side of the membrane, a gradient
is formed. One side contains more positive charges than the other side.
Once enough protons accumulate on one side of the membrane, another protein, known as ATP synthase, provides a channel for the protons to flow back across the membrane. As the protons flow through
the channel, the energy they release is captured to make ATP, the high-
energy molecule that can be used as cellular energy currency to help
the microbe accomplish many of the tasks it needs to survive. While
we have to eat other organisms to get chemical energy in the form of
ATP, many microorganisms use this simple and elegant system to convert sunlight into useful chemical energy. Like plants, such organisms
are known as photoautotrophs, meaning self-reliant light-capturing
organisms.

Holding pigments inside their basket structures has made opsins
proficient at capturing light, enabling humans to see many different
colors and archaea to generate their own energy. Yet long before opsins emerged, proteins known as cryptochromes were likely able to detect and react to light. Ancient cryptochromes found in our bacterial
ancestors help protect them from DNA damage caused by ultraviolet
(UV) light.[11] Such damage can be passed down as harmful mutations
to the next generation of bacteria. Cryptochromes are the bacteria's
clever way of using the same harmful UV light that causes damage to
DNA as the energy source to repair that very damage. Within each cryptochrome is a pigment known as a flavin, derived from vitamin B2.

Flavins are experts at absorbing blue and ultraviolet light. Within the cryptochrome, the light energy captured by the flavin pigment is used to scan the DNA for damage and to drive chemical reactions that repair the damage caused by UV light. The cryptochromes, in a sense, use the poison as the cure in an elegant system whereby more UV light triggers the action of more cryptochromes to find and repair mutations.[12]

We have our own version of cryptochromes, but over eons of evolution, our cryptochromes have lost the DNA repair mechanism found in ancient bacteria. Instead, we use our cryptochromes to regulate our circadian rhythm—the body's natural ability to adjust its daily sleep cycle by reacting to light cues. Residing in the retina and living side by side with the opsins we use to see color, cryptochromes are the ticking heart of our internal clock. Our cryptochromes receive light in the blue and violet range, using a similar flavin pigment. As they collect light, they move and wiggle in a coordinated dance that signals it is time to get up and move. We feel groggy on gloomy days when our cryptochromes don't receive enough light, and jet lag sets in when our cryptochromes are confused by a disrupted cycle of night and day.[13]

When plants emerged on the planet, they found a way to use cryptochromes to seek light and grow toward it. Cryptochromes serve as a plant's eyes. We may think of plants as nonmotile, anchored to the soil in which they are planted. But plants move with intention as they grow and branch, responding to environmental cues, most notability light. Cryptochromes on leaves and buds seek out light and tell the rest of the plant how to grow and branch to get the most sunlight. For a plant, which gets most of its energy from light, cryptochromes are indispensable to its survival. They also perceive the changing of the seasons. As the nights grow longer, cryptochromes sense the coming of autumn.

A TASTE OF HOME

In my father's newly upholstered Fiat, we left the brothers' upholstery shop and headed home to pick up my mom and my sister. On the way, my father and I retold the story of Mo to each other and shook our heads, smiling in disbelief over how the blind man was able to perceive what others could not see. The conversation changed when my mom and sister entered the car. They complimented the new seat covers. My mother praised my dad for deciding to get it done. On our way to my grandmother's apartment, we stopped at a bakery to pick up bread for dinner. As I got out of the car, I was instructed to buy a few more loaves of pita bread than usual. I didn't complain. Who doesn't love bread? For Egyptians, especially, bread has been an inseparable part of the culture for thousands of years. Bread has sustained life along the narrow Nile valley for about as long as wheat has been cultivated, and it has become synonymous with life itself. In fact, Egyptians seldom use the standard Arabic word for bread, *khobz*. Instead, we use the word *aish*, which means "life." The taste and smell of freshly baked Egyptian pita bread transcends age, social status, and economics. It is the taste of home.

Many years later, when the four of us had found a new home in the rural Midwest of the United States, the first thing we sought was the taste of familiar foods. The local supermarkets offered breads from all over the world: French baguettes, Italian loaves, Mexican tortillas, Jewish challah, and peasant ryes, but none could quench the longing for the bread we grew up eating—our *aish*. And it wasn't just the bread. Everything tasted different. Making food as a new immigrant is a challenge. You oscillate between the exhilaration that comes from trying new flavors and the disappointment of eating dishes that have familiar names, but just don't taste the same as they did back home. The Chinese buffet we frequented after church in northern Indiana was a brave new world of flavors and tastes that I had never experienced. The combination of textures, the color of each flavor, and the intensity of the spices were a departure from the familiar Middle Eastern palate I grew up with. With each new dish, my taste buds either rejoiced, recoiled, or were simply confused.

Human tongues contain roughly five thousand taste buds, each living for about ten days before being recycled.[14] Our taste buds inhabit the surface of the tongue as well as parts of our cheeks and throat, extending all the way toward the inside of our esophagus. Shaped like a mushroom, each protruding tastebud is composed of hundreds of supporting cells and gustatory cells. Supporting cells provide a rigid base and a strong structure for gustatory cells, which work as the collection hubs for taste molecules. The gustatory cells are long, reaching out like flower petals, with hair-like structures that help gather taste molecules as they swim over our tongues. On the surface of the gustatory cells live hundreds of thousands of taste receptors—the proteins at the frontline of our ability to taste food.

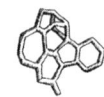

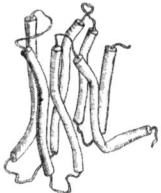

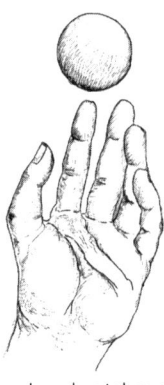

Just as a hand catches a ball, taste receptors, like this bitter taste-receptor protein (PDB code 7XP6), catch taste molecules like this bitter molecule known as strychnine. The binding of a molecule to the taste-receptor protein triggers a unique taste sensation.

All the tastes, new and old, strange and familiar, belong to five basic tastes: sweet, salt, sour, bitter, and umami (savory). Each taste is mediated by one of five types of taste-receptor proteins that populate the surface of our tongues. Most taste-receptor proteins, particularly those we use to taste sweet, bitter, and savory flavors, look a lot like opsins, with a similar basket shape anchored onto the surface of our gustatory cells. But whereas opsins use an embedded pigment to receive light, taste-receptor proteins attract and bind taste molecules, using a type of molecular Velcro. Each taste-receptor basket has a slightly different receiving end, into which different taste molecules fit. Like a hand with fingers pointing up, ready to catch a falling ball, taste receptors "catch" the molecules in finger-like "rods" or "helixes." This class of receptors has seven of these rods, which form the walls of the protein's basket structure.[15]

Many taste receptors have extra appendages that reach high above the membrane of the gustatory cells. With their long stem and "mouth," which grabs onto taste molecules as they float across our tongues, these appendages resemble a Venus flytrap plant. Taste receptors can even work in pairs; anchored together, they communicate by nudging each other whenever one grabs onto a certain taste molecule. The nudge from one taste receptor to another elicits a response that is detected by proteins on the inside of the gustatory cells. Those proteins in turn cause the efflux of large amounts of positively charged calcium ions from stores inside the cell. This generates an electric pulse that gets carried by nerve cells to the area of the brain responsible for our sense of taste. The key to sensing different flavors is shape complementarity—how a flavor molecule fits within the specific shape of a receptor like a 3D puzzle piece. Sugars will predominantly fit within the shape of the sweet receptors, while acids from lemons or vinegar will fit into the shape of the sour receptors.[16] Each receptor sends an electric pulse to the part of the brain responsible for its own flavor, combining to give the collective experience of each unique dish, familiar or foreign.

APPLES AND ORANGES

One rainy winter day in 1986, my aunt and uncle returned to Egypt from a trip to West Germany. They came back loaded with gifts—little trinkets and toys, giant bars of hazelnut chocolate, and various Bavarian souvenirs. They also brought with them a handful of chestnuts, squares of licorice and marzipan, and a few large red apples. These were all flavors I had not experienced before. I didn't care much for the licorice or the marzipan, but when my mom began roasting the chestnuts, I was immediately enchanted by a scent like nothing I had ever experienced. We cracked open the singed shells to reveal the soft beige insides, insulated by a fluffy cover. "Abu farwa," my dad exclaimed, using the Arabic word for a man wearing a fur coat. The first bite was a buttery sweet and nutty flavor that was as foreign to me as the Bavarian countryside with its snow-covered rooftops and large evergreens. Then came the apple, which I saved for last. It was shiny, almost waxy, and deeply red. I took little bites, relishing the flavor over my tongue and trying to hold on to the memory of the taste.

Most foods rarely consist of just one taste molecule. The taste of an apple comes from a variety of molecules, some of which trigger sweet receptors, some sour, and some bitter. Different varieties of apples have different ratios of these molecules, which give them their unique tastes. The collective "apple experience" comes from the combined signals of the five basic types of taste receptors—sweet, salty, sour, bitter, and umami—as each one is triggered to a different extent. Moreover, our taste receptors are inherently nonspecific. Flavor molecules swimming over our tongues find their way into the loose grip of a taste receptor through shape complementarity. A molecule landing in the receptor is less like a key fitting into a lock and more like a loose handshake, where the receptor twists to adjust to the shape of the taste molecule. This allows a handful of receptors to grab onto a large number of taste molecules as long as they have somewhat similar shapes. Some molecules have an easy time fitting in and bind tightly to the receptor, eliciting a strong response. Others require significant adjustment by the receptor and bind loosely, triggering a more subtle response. The combined triggering of the different taste receptors at different levels relays the collective experience to the brain, which in turn perceives the experience as a unique flavor profile. In the same way that only three opsins help us see the entire range of the rainbow of colors, our five types of taste receptors work together to relay the complexity of a nearly endless number of flavor profiles.

While the looseness of the taste receptors allows us to perceive a wide range of flavors, it also means that our receptors can be tricked. Our sweet receptors will bind to sugars, but they will also bind to molecules that look like sugars. The artificial sweetener in Nutra-Sweet, known as aspartame, is the perfect example of how our senses can be fooled. Aspartame is not a sugar, but it binds very tightly to the sweet receptors. In fact, aspartame binds so tightly that it triggers a sensation of sweetness nearly two hundred times stronger than an equivalent amount of table sugar.[17] This allows us to use a tiny bit of aspartame in soft drinks, tea, or coffee without having to consume as many calories.

As important as taste receptors are to our ability to taste our food, much of our flavor experience is connected to our sense of smell. Indeed, with our limited number of taste receptors, we rely heavily on our sense of smell to get the full experience when consuming our favorite meals. The complexity of the flavors in food comes from the interplay between taste and smell receptors. When blindfolded, a person smelling a pear and eating a slice of an apple may think that they are eating a pear instead of an apple. The scent from the pear dominates the perception of what fruit they are eating. Or consider a time when you couldn't smell much of anything. When you catch a cold, your nasal cavity typically swells, preventing your olfactory receptors from receiving scent molecules from the air. This is why nothing tastes quite right when your nose is stuffy.

With the same characteristic basket shapes as opsins and taste receptors, smell receptors, also known as "olfactory receptors," are constantly collecting scent molecules from our surroundings so that we can decipher even the slightest changes in our environment. Olfactory receptors are also made of the seven-helix (rod) framework, but unlike sight and taste, which involve only a handful of different receptors, our sense of smell employs an immense number of different olfactory receptors. Nearly a thousand different genes code for these receptors. Over time, we lost the ability to turn many of those genes into proteins, developing nearly six hundred "pseudogenes" that never produce functioning receptors.[18] Still, we have about four hundred different olfactory receptor proteins that line our nasal cavities and act as a front door to the system that allows us to smell our world.

At the center of our sense of smell is a group of cells that form a structure known as a glomerulus (plural: glomeruli). This collection of cells acts as the relay station in the process of translating the chemicals floating in the air to the perception of smell.[19] Glomeruli reside on the bottom side of our brains, just above our noses. From each glomerulus, long wirelike nerve cells extend all the way down into our nasal cavities. On the surface of these nerve cells, olfactory receptors are displayed, each with a different shape able to grab onto different scent molecules. Each glomerulus receives signals from one type of olfactory receptor. Numerous copies of each unique olfactory receptor are displayed on 10 to 20 million olfactory nerve cells.[20] These allow us to decipher an estimated one trillion different scents.[21]

As astonishing as this seems, our smell perception pales in comparison to a dog's legendary sense of smell. Dogs have roughly 1,500 different kinds of olfactory receptors, with over 250 million copies displayed inside their noses—many times more than what we display in our nasal cavity.[22] One can only imagine how many more scents a dog can experience. It's no surprise that dogs have been trained to sniff out explosives, find victims trapped under the rubble of collapsed buildings, and even detect cancers in humans—all with a sensitivity rivaling that of our most sophisticated diagnostic machines.[23]

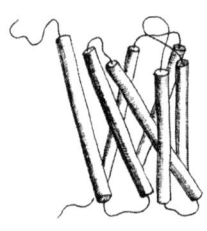

Dogs have an exquisite sense of smell, all due to nearly 1,500 different olfactory receptors similar to the one pictured here (PDB code 8F76).

It may come as a surprise that animals dedicate such a large number of proteins to detecting scents, compared to the few dedicated to sight and taste. But the sense of smell is thought to be the most ancient of all senses, carved deep into our primordial past and carried through billions of years in the genes we received from our ancestors. Before life made it to land, living in a primordial ocean required the ability to "smell" underwater. Long before humans or any multicellular organisms existed, tiny bacteria developed a rudimentary sense of smell to help them locate and swim toward nutrients while avoiding dangerous or toxic molecules. To detect molecules in their environment, bacteria use receptor proteins just below the outer shell of their single-celled bodies. Bacteria, like humans, dedicate a large portion of their genomes to making receptors that help them navigate their environment, but unlike our basket-shaped receptors, bacterial receptors are shaped more like a Venus flytrap, with two flaps connected by a hinge. One such receptor is known as glucose binding protein.[24] It latches onto the sugar glucose and snaps shut the way the flytrap latches onto its prey. The change in the shape of the protein from open to closed is detected by nearby proteins known as transducers, which immediately begin working together to bring the sugar into the cell to be broken down for energy.

This relay process between the receptor and the transducer proteins is not only important for capturing food in the environment, but is also closely linked to how bacteria move. Recall that bacteria are propelled through the environment by flagella composed of long protein filaments anchored to one end of the bacterial cell and attached to a tiny molecular motor, which is also made of proteins. Much like a boat propeller, the motor at the base of the flagellum powers the corkscrew to move the cell forward. When a nutrient like glucose is received by the receptors at one end of the cell, the transducer proteins send chemical signals to jump-start the motor, which turns the flagellum, moving the bacteria closer to the source of the sugar in what scientists call a "run."[25] The opposite happens when a toxin is detected by one of the many toxin receptors on the cell surface. In this case, the flagellum stops turning, allowing the bacterium to tumble and change its orientation before the motor starts up again, pushing the cell away from danger. By alternating cycles of run and tumble, the bacterium can swim toward food and away from harm.[26] It's astounding to think about the scale of a sensor like a glucose binding protein relative to the size of the bacteria. The entire protein is millions of times smaller than the bacterial cell, yet when glucose is captured, a minuscule hinge-bending motion triggers a set of chemical reactions that ultimately decide the direction in which the entire cell moves. Like a rudder on a massive ship, a bacterium uses the protein receptors on the cell surface to navigate its environment, "sniffing out" what's beneficial and swimming toward it, or moving away from what's harmful.

A similar run-and-tumble mechanism is found in human sperm cells. Olfactory receptors, much like the ones found in our nasal cavities, reside on the surface of sperm, guiding their journey as they follow the "scent" of an unfertilized egg.[27] And like their bacterial ancestors, sperm use powerful flagella to propel themselves forward. But unlike the corkscrew motion of the bacterial flagellum, the sperm flagellum is built more like a whip and sends the sperm forward in a slithering motion. This basic strategy for following a chemical, which started with bacterial navigation, has persisted through evolution, taking slightly different forms but keeping the same framework intact. Salmon are a macroscopic expansion of this framework. They use olfactory receptors to detect tiny molecular signatures that can guide them to the streams where they first spawned.[28] Following the scent, salmon retrace their own steps up the stream and continue a pattern of chemical detection and navigation that was pioneered by bacteria millions of years before any fish existed.

Many of our human olfactory receptors are remnants of an ancient world of scents, part of a system so primal that olfactory receptors develop our perceptions of the world even before we are born. In fact, our sense of smell develops in the womb.[29] As we breathe in our mother's amniotic fluid, we become familiar with her scent—a complex mixture of her genetic makeup and remnants of the molecules in the food she consumed. Recent experiments show that newborns tend to be attracted

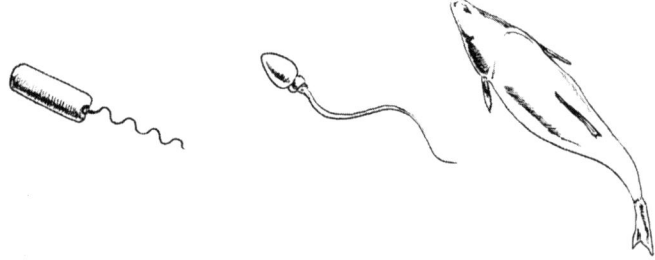

Finding our way using chemical signals is a universal mechanism that spans multiple kingdoms. Bacteria use binding proteins to direct their flagellar motors toward nutrients, human sperm use olfactory receptors to track the scent of an unfertilized egg, and salmon use olfactory receptors to trace their way back to the stream where they spawned.

to the foods consumed by their mothers while they were pregnant and breastfeeding.[30] In one experiment, scientists divided pregnant women into two groups, with one group consuming a lot of carrot juice in the third trimester or while breastfeeding. The babies of women in this group were found to be more enthusiastic about foods containing carrots, while the babies of women in the other group showed no particular preference. Another experiment conducted on rats showed that consuming large amounts of "junk food" during pregnancy made the offspring more drawn to junk food and increased their chances of developing obesity.[31] These experiments demonstrate the importance of taste and smell during development.

Our preference for the foods our mothers consumed while pregnant and breastfeeding likely provides a protective advantage in our search for safer food sources by drawing on our very first memories— what our mothers ate. It's perhaps why we tend to gravitate toward comfort foods, the scents and flavors we consumed in our childhood homes. Such foods represent the safety of what's familiar. Moreover, long before our ability to see is fully developed as babies, we recognize our mother by her scent. As we grow into toddlers, our noses help us learn about places and people. Eventually, we learn to rely on our sight in most aspects of daily life, but deep within our primordial core, familiar scents are woven into our molecular fabric. In many ways, our awakening to the world begins long before we open our eyes or take our first breath.

Olfaction is perhaps even more crucial for the survival of our animal relatives. While humans can start a fire or turn on a light, most animals spend half of their lives in the darkness of night and must rely on a keen sense of smell to navigate their environment, find food, or avoid danger. Animals, whether on land or in the ocean, use olfactory receptors to grab onto scent molecules floating in the air or in water and translate those molecules into decipherable signals. These signals tell the animal if there is food, danger, or a mate ahead, so it can act accordingly. Grabbing the tiniest invisible molecules from the air is not an easy task, and olfactory receptors must be extremely sensitive. Even more challenging is sorting out the immense number of odor molecules that animals are immersed in at any given time, a process that involves cooperation and coordination. Like keys on a piano, which produce a variety of sounds depending on which are played together and when, the collective effort of olfactory receptors paints a more complex and nuanced picture of the animal's surroundings than the images we see with our own eyes on the sunniest of days.

The large number of olfactory receptors humans have carried through our evolution and the ancient nature of the sense of smell itself may be why scents often have such a profound effect on our emotional states. The sense of smell is deeply entwined with our memories. Certain scents can elicit strong emotional responses that trigger joy, disgust, sadness, and even pain. Scents can also bring back memories from a long-lost past by triggering olfactory flashbacks. Ten years after moving to the United States, I took my first trip back to Egypt. I was no longer a nerdy teenager; I was in my twenties, confident and accompanied by my wife. The sun was setting over the bustling Cairo sky as we rode in a cab from the airport. The warm air flowing through the rolled-down windows carried the strong and ever so familiar scent of Egyptian bread from a nearby bakery, one I had not smelled in nearly ten years. My confidence was wiped away, and I was overcome by a flood of emotions, as memories of a faded past came back all at once. The mere scent of bread triggered a profoundly moving experience. It brought back a flood of memories from my childhood home, dinners with extended family, summer vacations and chilly winter nights, the stories and the chatter, the noise from the neighborhood where I grew up, and even distinct images of the pattern on the curtains hanging over the tall balcony doors. I was experiencing an olfactory flashback, a primordial connection to a deep and complex mixture of emotions and family ties from a distant past. It was more than just a reminder of the bread that had nourished my body—the scent had transported me back in time, to feel once again the love of a family and yearning for a long-lost home.

Genetically speaking, most humans have nearly identical sensory receptors. But even with very similar receptors, we process scents in our environments very differently. The same smell of bread that elicited nostalgia in me may trigger a completely different feeling in someone else. The receptors may be our windows to the same world, but each of us uses them in a deeply personal way. Even tiny changes in the genetic makeup of these receptors can have a significant influence on our food preferences. For instance, some of us find spinach and broccoli to be tasty, while others can't stand them. Studying these two groups of people has revealed a connection between how we perceive the taste of these vegetables and our ability to taste a chemical known as propylthiouracil. People who are sensitive to the bitterness of propylthiouracil found in vegetables find spinach and broccoli to be extremely bitter. The underlying differences between the two groups result from differences at the genetic level, specifically in the gene coding for a bitter-taste receptor protein known as TAS2R38.[32] Tiny changes in the makeup of the TAS2R38 protein make some of us abhor the taste, while others love it.

KNOW THYSELF

As we move through our surroundings, we grab molecular cues in sights and flavors, scents and sounds, and process them through the complex maze of proteins that inhabit the billions of cells in our nervous system. We associate foods and scents with events and emotions. Collectively, these experiences form our unique worldview, ultimately defining our identity as humans. While sight, smell, and taste occupy different spaces in our human consciousness, they each provide a different dimension to our collective experience of awareness. On the molecular level, they share the same basic mechanics. In fact, opsins, olfactory receptors, and most taste receptors belong to a single family of proteins and share the familiar basket shape.

Another common feature of these receptors is how they transmit a signal once they catch a molecule in their basket. At the heart of their signaling mechanism is a small companion protein known as a G-protein. This protein is always bound to the receptor until a scent or taste molecule falls into its basket. At that point, the G-protein leaves the receptor and signals to the rest of the cell that a molecule has been captured. This dependence on the G-protein has earned these receptors the name "G-protein coupled receptors," or GPCRs for short.[33] Once the G-protein is activated, a cascade of chemical reactions is initiated as more proteins bump into others, propagating and amplifying the signal at each step. The result is an electrical pulse that travels through the nerves all the way to the brain, where other proteins decipher the signal as a unique smell, taste, or color.

GPCRs are not merely our gateway to the world; they are also the key to self-awareness. Many hormones that regulate our bodily functions relay their signals and trigger major physiological changes through the action of GPCRs. When animals feel fear or sense a threat, the hormone adrenaline is released from the adrenal glands, and within seconds, the bloodstream is flooded with adrenaline, triggering the fight or flight response and that ever so familiar adrenaline rush. Adrenaline signals to our body to respond to danger by finding and sticking to special GPCRs known as adrenergic receptors. These proteins live on the surface of many different parts of our bodies, including the heart, lungs, and fat and muscle cells. Once adrenaline binds to the adrenergic receptors on the surface of our heart, it begins to beat faster and stronger, sending blood and nutrients to our muscles. Our lungs expand, and we breathe faster to provide more oxygen to the muscles. They in turn, upon receiving adrenaline through their own adrenergic receptors, begin to tap into stored sugar and fat reserves for much-needed energy to fight the imminent danger or to dash out of harm's way.

While some GPCRs, like the adrenergic receptors, relay an excitatory effect, other GPCRs trigger the opposite outcome.[34] Long-distance runners are familiar with the term "runner's high." This is when endorphins are released into the blood in response to the stress of exhaustion. The same endorphins produce the euphoria we experience after a very spicy meal, as long as we can withstand the initial pain. These endorphins are released from the pituitary gland in the brain, find their way to specialized GPCRs in the nervous system, and trigger a sense of elation and pain relief. The same GPCRs that bind to endorphins also bind to opioids like morphine and heroin, but these plant-derived molecules have a greater effect on our bodies. Unlike the mild effect of endorphins, opioids cause intense euphoria and can lead to addiction. Even though in both cases the same GPCRs are activated, opioids can produce an effect that is over a hundred times stronger than that of our natural endorphins.

With GPCRs, typically, more activation leads to a greater response. For example, more adrenaline will turn on a stronger sense of fight or flight. But because the structure of a GPCR is moldable and malleable, and each GPCR can accommodate one of several different molecules with slightly different structures, the activation of the same GPCR can sometimes result in very different outcomes. Consider the GPCRs responsible for serotonin signals. Serotonin is commonly known as the happiness molecule. It affects our mood, energy, and sleep. Problems with low levels of serotonin or dysfunction of the serotonin receptors often lead to depression.[35] This is because when serotonin binds to its receptor, physiological changes in our brains trigger elation and happiness. The synthetic drug known as LSD also binds to the serotonin receptors, but has a very different effect.[36] LSD causes intense hallucinations and sensory disruptions. Why the difference? Serotonin and LSD tweak the basket structure of the receptor in slightly different ways, leading to different chemical reactions downstream and profoundly different experiences. As it turns out, even subtle changes in the structure of the GPCR at the nanometer scale (1 billionth of a meter) can be detected by surrounding proteins, leading to two very different outcomes.[37] So small tweaks in the way that GPCRs change their shapes in response to the different molecules they receive can have a profound effect on how the entire body reacts.

What complicates matters even more is that the serotonin receptor is only one of thousands of GPCRs that inhabit our nervous system. It is the combined effect of the actions of these amazing proteins and other types of receptors that ultimately gives us the wide range of feelings we experience. Dopamine receptors, for example, are a set of GPCRs involved in regulating memory, attention, cognition, and impulse control. Addictive behaviors are known to trigger the release of dopamine, which enhances pleasure and promotes more of the addictive behavior. Smoking, drinking, and even doom scrolling all activate dopamine, which is why it can be so difficult to stop these behaviors.

Oxytocin receptors are another type of GPCRs that are tightly connected to our emotions.[38] They play an essential role in creating the bond between mother and infant and may also be important in promoting other social and emotional interactions. While it's still not completely clear how oxytocin affects these behaviors, there's evidence that dysfunction in the oxytocin receptor may be associated with postpartum depression, anorexia nervosa, and social anxiety. The receptor may also be essential to the way we empathize with others and how we interact within a social group. Pain, joy, fear, anxiety, depression, ecstasy, remorse, and even love arise not from the action of a single receptor but from the complex crosstalk of a vast network of receptors and thousands of other surrounding proteins. The complexity of the web of protein receptors and their intertwined signals provides us with an incredibly wide range of emotions, some we don't even have words to describe, yet so many that we share as part of our human experience.

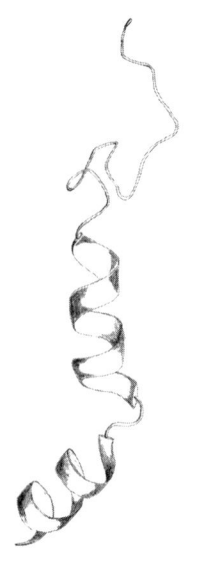

Orexin (PDB coed: 1R02) is a small protein that binds to a GPCR known as the orexin receptor. Orexin signals our bodies to wake up by activating the orexin receptor, which then triggers physiological changes that help us get moving.

Even in our sleep, receptors work together to allow us to experience elaborate dreams in which we see, smell, walk, and talk. We can move through our dreams without our muscles firing. This is a type of paralysis that disconnects our thoughts from our physical actions to protect us from hurting ourselves until we are ready to wake up. At that point, a small protein known as orexin is released from the brain to signal wakefulness. Orexin binds to the (you guessed it!) orexin receptor. Found deep inside the brain in the hypo-thalamus, the region responsible for controlling our vital functions, the orexin receptor is a GPCR with the characteristic basket shape. It is one of a handful of receptors that play an important role in recon-necting our brains to our bodies so that we can successfully get out of bed. The production of orexin must be tightly regulated: it has to be turned off during sleep and turned on when it's time to wake up. Prob-lems with regulating orexin, as one can imagine, often lead to sleep disorders like narcolepsy or sleepwalking. In some cases, it can even lead to a type of sleep paralysis, where a person is awake but tempo-rarily paralyzed. Many individuals experiencing sleep paralysis report seeing alien shadows, hallucinations, and out-of-body experiences. While the exact mechanism of this condition is not known, it is most likely due to asynchrony between the different receptors responsible for awakening—a mistiming of the handshakes between molecules like orexin and their receptor proteins. Studying the structure of the orexin receptor and how it interacts with orexin has led to advances in our understanding of the cycle of sleep and wakefulness on the molec-ular level. These studies have also led to the design of a therapeutic molecule known as suvorexant, which blocks orexin from binding to its receptor. Suvorexant is now used to inhibit the wakefulness effects of orexin, providing much-needed sleep to those suffering from chronic insomnia.

The sheer number of GPCRs in our bodies and the complexity of the different shapes that they create make them the focus of intense re-search. Not only are GPCRs important in depression and insomnia, but they also act as gatekeepers for the onset and progression of disease. GPCRs control when we store or break down our food, how we excrete toxins and waste products, and how we respond to pain. This makes them essential players in diseases like diabetes, hypertension, hypercho-lesterolemia (high blood cholesterol), chronic pain, and even in the onset and spread of many cancers. It is no surprise that nearly 40 percent of all the drugs currently on the market target GPCRs in one way or another.[39] Members of the diverse family of GPCRs are at the heart of our ability to sense the world through taste, sight, and smell, and are key to communication between the different parts of our bodies. In many ways, GPCRs are our gateway to awareness, both in perceiving the outside world, as well as in connecting the different parts of our bodies to each other through the molecular language of chemistry. The molecular Velcro of GPCRs enables them to gather precise chemical information from scents, tastes, hormones, and drugs. Each molecule that finds its way into the basket structure of a GPCR is translated into a sense, a feeling, or a memory that elicits a unique physiological response, ultimately molding our sense of who we are in the world.

Yet GPCRs are not the only type of receptors we use to sense the world around us. For example, the receptors on our skin for touch, heat, and cold sensations are proteins of a different kind. Instead of using a G-protein to facilitate their signals, many touch receptors facilitate the movement of ions across the cell membrane to trigger an electric potential.[40] These are small voltage changes that propagate through our nerves, making their way to our brain through the spinal cord. These receptors send their signals using the extensive network of nerve "wires" that run through our bodies. The electric signals created by an activated touch receptor can travel through our bodies at nearly 130 miles per hour.[41] But even such a high speed can be too slow to prevent injury. If you touch a hot pan on the stove, by the time the signal reaches the brain and a decision is made to move away, extensive tissue damage may already have occurred. At times like this, a critical shortcut is activated: the hot pan sensation will travel only as far as the spinal cord, which will send a quick signal back to our hands to pull away. This type of reflex happens literally before you know it.

WHAT COLOR IS NORTH?

Every fall, as the trees change colors, a small bird on the Gibraltar Peninsula prepares to make a journey. She fills her belly with worms, collecting energy for a long trip hundreds of miles north across the Iberian Peninsula to her summer home in Central Europe. There, this European robin will nest, breed, and raise the next generation of robins. Without the complicated switches and dials found in a jumbo jet cockpit, this robin, like so many other migratory birds, finds her way through the sky with incredible ease. The exquisite ability of migratory birds to navigate the globe has been a source of fascination for centuries.

As early as two thousand years ago, humans learned to use magnetic metals to navigate land and sea. The earliest compass consisted of a pin made of iron or another magnetic metal floating in a bowl of water. As the pin rotated freely over the water's surface, it would align itself with the magnetic pull of the Earth and with the Earth's north-south axis. Today, more complex instruments are used on planes and ships, but the basic idea of orienting to the Earth's magnetic field has persisted. For decades, birds too were thought to use a magnetic metal like iron to navigate the skies, much like the way we use a compass. But more recent evidence suggests that at the heart of the bird's navigation skills is a protein: a cryptochrome. Like all proteins, cryptochromes are made of non-magnetic elements like carbon, oxygen, hydrogen, and nitrogen, none of which can align themselves with a magnet.[42] Yet deep within each one of these elements are miniscule electrons. Under the right conditions, electrons will align with a magnet the way the pin of a compass points north. But often, electrons come in packages of two, and when they do, the paired electrons align in different directions and "cancel each other out." Since all the electrons in the elements that make up proteins are paired, proteins are rendered non-magnetic and unable to distinguish north from south.

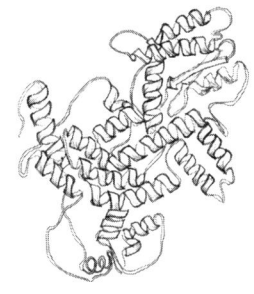

European robins are among birds that use a special protein to find their way through the sky. The bird protein known as cryptochrome (PDB code 1QNF) has the ability to align a single electron within its core to the magnetic field of the Earth. This is key to how the bird finds its way on a long journey. When the bird looks at the sky, north looks different from all the other directions.

To overcome this, cryptochromes employ a clever trick, which takes advantage of their location in the eyes of the bird. As sunlight enters the bird's eye and collides with cryptochromes, the light energy is used by the protein to tear apart a pair of electrons, creating a molecule with an unpaired electron. This "lone electron" acts like the needle of a compass, guiding the robin as she flies north to her summer home.

What's incredible about this system is how the birds process the information from the tiny lone electron inside the cryptochrome.[43] Cryptochromes provide a picture of the sky that travels through the optic nerve and gives the bird a vision of what "north" looks like. To us, the sky may look clear or cloudy, blue or gray, but a bird can see more than just clouds. As the cryptochrome in her eyes splits a pair of electrons, the bird sees a unique color that it calls north. It seems there's far more to a bird's-eye view, with colors that we can only dream of perceiving.

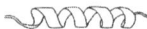

That a bird that can see the color north or an octopus can see with its skin seems stranger than science fiction. But everywhere we look, we find examples of extraordinary senses that we can hardly imagine with our limited sphere of perception. We find bats navigating dark caves with sonar and whales detecting sound waves far outside our hearing range.[44] We see how sharks can detect electromagnetic waves emitted by a fish hiding in the sand, and how pit vipers can build a heatmap of a scurrying mouse in total darkness.[45] We find Komodo dragons tasting the air for clues of prey and elephants sensing earthquakes from miles away.[46] So many remarkable senses have yet to be studied, and many more are waiting to be discovered. We are only beginning to understand the vast world of senses around us, and how small a slice of it we humans perceive. But one thing we know for sure: all these senses are mediated on the molecular level by the interplay of proteins.

Most organisms whose senses surpass our own have inhabited the Earth much longer than we have. In fact, if the entire time of life on

Earth was just one calendar year, humans would not appear until the evening of December 31. To the octopus, we are the newcomers, and with our hair-covered dry skin and only four bony limbs, we might as well be the aliens. It has long been tempting for humans to think of evolution as a progression that produces more "advanced" species, and to place ourselves as the pinnacle of life, endowed with the most important traits of all living things. Indeed, humans possess some amazing abilities that have helped our species spread to all corners of the planet. But evolution is more accurately described as slow change over time through adaptation.[47] While complexity can emerge and expand over time, that does not necessarily mean that evolution creates "better" creatures—only that they are more adapted to a given environment. While bacteria may seem primitive to us, they have all the components necessary to survive where they live. They don't need our complex brains any more than a snake needs legs. Humans, as evolutionary "newborns" on this planet, may still have much to learn from them—and from all the other elders who inhabited and guarded the Earth for billions of years before we arrived.

4

TAKING FORM

SHAHIR S. RIZK AND MAGGIE M. FINK

LINUS PAULING WAS RESTLESS. The forty-seven-year-old scientist, who was often sickly and too weak to work, had fallen ill with a cold and was supposed to stay in bed for three days. It was the year 1948, and Pauling, bored with reading detective novels, turned his attention to a question that had been on his mind for months: what do proteins look like? At the time, very little was known about protein structure. Pauling, a chemistry professor at the California Institute of Technology who was spending the year as a visiting professor at Oxford University, was an accomplished chemist, but even for him, protein structures remained elusive. Many scientists had worked with complicated machinery and devised complex formulas to try to predict what shapes proteins took on, yet very little was clear. Pauling took out a piece of paper and scribbled down a few chemical structures, sketching out what was known about proteins. He drew carbons and oxygens and connected them to nitrogens and hydrogens in the way he knew that amino acids, the building blocks of proteins, were arranged. He worked until he had a repeating pattern of chemicals. The letters formed a long line across the page, which he cut out into one long strip. He then took the strip of paper and twisted it around itself, carefully lined up the atoms that attracted each other, and found that the strip twisted into a spiral. That February morning in Oxford, Pauling had stumbled upon one of the most fundamental shapes found in all living things. He had discovered the alpha helix, a shape that would later become synonymous with the language of proteins and the building blocks of life itself.[1]

Long before Pauling pondered the shape of proteins, scientists worldwide had tried to imagine the basic structures of the fundamental building blocks of our world. In fact, questions concerning what we are made of are as old as humanity itself. Every culture has a creation story, often one that combines "stuff" with "spirit" to explain the foundation of our being. What that "stuff" is made of, and what shapes it takes, have occupied the human mind for millennia. Greek philosophers believed that all matter belonged to one of four fundamental "elements": earth, fire, water, and air. Among these important philosophers, Democritus believed in the concept of an atom—the fundamental building block of every element, something so small that it could not be broken down into smaller parts. But without experimental evidence, the idea of an atom could be neither proven nor dismissed.

Around the eighth century CE, Arab scientists began to develop the field of alchemy. Initially rooted in magic and myths, alchemy eventually evolved into a proper science and became the precursor to modern chemistry (the word "chemistry" is even derived from the Arabic word *al-kīmiyā*). One of the most prominent figures in chemistry during the Arab golden age of science, Jabir ibn Hayyan, is credited with the development of many chemical processes still used today for dying clothes, tanning leather, and protecting metals from rust.[2] He is also credited with developing several techniques for separating and purifying chemical compounds involving crystallization and filtration. Ibn Hayyan is said to have developed distillation, an essential step in making many alcoholic beverages and even in the production of medicines. Distillation involves boiling complex mixtures, then cooling their vapors to condense them into liquids that can be separated into different components. Using distillation, Arab scientists could separate a complex mixture, like wine, into its individual components such as alcohol, sugar, acid, and water. These compounds, which could not be separated any further, were considered "pure substances." Yet what those substances were made of remained elusive.

Building on the work of Arab scientists, a French husband-and-wife team, Marie-Anne Paulze and Antoine Lavoisier, discovered around 1770 that these so-called pure substances could be broken down into smaller components when subjected to chemical reactions.[3] Water, for example, was found to break down into oxygen gas and hydrogen gas. No matter how much they tried, the gases could not be broken down into simpler components, so the couple concluded that they must be "elements"—substances that cannot be broken down further or made by combining other components. The work of Paulze and Lavoisier brought back the Greek concept of the indivisible atom and revised the idea of pure elements. Before his execution by the guillotine at the height of the French Revolution, Lavoisier was among a group of scientists who had dismissed the idea that earth, fire, water, and air were the fundamental elements in favor of a revolution in chemistry that described more than fifty elements. Today we know of no fewer than 118 elements, each of which occupies a slot on the modern periodic table.[4]

The discovery of the elements established that these indivisible atoms could come together to form molecules or compounds. Two oxygen atoms combine to form the molecules of the oxygen gas we breathe; two hydrogens come together to make hydrogen gas, the fuel of stars and the most abundant gas in the universe; and two atoms of oxygen combine with one carbon to make molecules of carbon dioxide, which plants and algae consume to build food stores. Yet for decades, scientists didn't know how to describe the shapes of molecules, even molecules as simple as water.

The search for how elements came together to form compounds and what shapes they adopted continued into the twentieth century. There were hints that the shape of molecules affected how they behaved and how they reacted and interacted with other molecules. Questions such as why water is a liquid at room temperature when methane is a gas, and why alcohol mixes with water but oil separates from it, would surely be answered if the shapes of these molecules were better deciphered. Furthermore, knowledge of the shapes of these simple molecules could help inform our understanding of our own biology and how biomolecules function.

As scientists began to dig deeper into the question, the focus turned to the chemical bond, the handshake between any two atoms within a molecule. In a compound like table salt, made of sodium and chlorine, the bond could be explained by an attraction between a positively charged sodium (having lost one electron) and a negatively charged chloride (having gained an electron). The attraction between the opposite charges of sodium and chloride, known as ionic bonding, extends in three dimensions and forms the grains of salt we add to our food. But the bonds between the oxygen and hydrogen in something as simple as water could not be explained by attraction between opposite charges. After all, hydrogen and oxygen do not lose or gain electrons when they combine to make water.

In the early twentieth century, the concept of a "covalent bond," in which electrons can be shared between two atoms, emerged just as the field of quantum physics, led by scientists like Niels Bohr and Erwin Schrödinger, was getting its start. Quantum physicists used complex mathematics to describe how two hydrogen atoms come together to form a bond in one molecule of hydrogen gas. Their ideas on how electrons are shared and how they form discrete configurations were groundbreaking, but the math was so complicated that it was very difficult to apply to anything more complex than hydrogen gas. At the time, the shapes of only a few molecules had been deciphered. It was then that Linus Pauling came to the scene, taking interest in how atoms formed bonds. In 1939, Pauling, still an aspiring young chemist, published his transformative work *The Nature of the Chemical Bond*. In it, he challenged old ideas about how elements "hooked" together to form molecules and proposed new ones based in quantum physics.[5] With Pauling's simple yet powerful ideas, molecules began to take shape in the imagination of chemists. His work explained the geometries of many molecules, including water and methane.

In many ways, Linus Pauling's work changed the game. He came up with his own rules for how individual elements shared electrons within bonds—rules that not only helped explain already known structures, but also could predict with great accuracy the bonding properties and geometric arrangements of more molecules than ever before. His work, while deeply rooted in theory and backed up by laboratory observations, was wildly creative and imaginative. Most notably, his work was not bogged down by complex math. As a result, even a novice chemistry student could use it to deduce the structures of a vast array of molecules. While the other great chemists saw the world of molecules strictly through the lens of quantum physics, Pauling saw atoms come together like puzzle pieces, each with its own unique edges and precise ways in which it could fit with others to form the molecules we encounter every day.

By the time Pauling was sitting in bed, sick, and contemplating the nature of protein structures, he was already an accomplished scientist. His famous book had become the authoritative work on how individual elements come together to form complex molecules. It was a huge hit and was used as the chemistry book of choice at many universities. Pauling, the chemist, had gained the respect of scientists worldwide and become the go-to scientist when it came to chemical bonds. His most notable contribution to understanding bonding was his observation that some elements can be more selfish toward electrons than others. In a covalent bond, where two different elements share electrons, Pauling suggested that the electrons are seldom shared equally. He imagined two atoms engaging in a tug of war over electrons in a bond, each element trying to pull the electrons closer to itself. In a bond where the two elements are the same, like two hydrogens or two oxygens, the "tugging forces" are equal, and the electrons are shared equally. But in a bond between two different atoms, like oxygen and hydrogen, the electrons are not shared equally. Oxygen, being more selfish than hydrogen, pulls the electrons closer to itself. This selfishness he called "electronegativity." Oxygen, being more electronegative than hydrogen, pulls the electrons closer to itself, leaving the positively charged nucleus of the hydrogen somewhat exposed.

This slightly uneven distribution of the shared electrons in a bond has huge consequences for how the molecule behaves. Since electrons carry a negative charge, the oxygens in water carry a slightly more negative charge, and the hydrogens are left with a slight positive charge. This unevenness creates two oppositely charged sides within the same molecule. As a result, water tends to have a slightly negative side where the oxygen is hogging the electrons and a slightly positive side where the hydrogens are electron deficient. And so, water is known as a polar molecule. Many other molecules where uneven sharing takes place, like ammonia, alcohol, and many acids, are also polar, and this makes them dissolve readily in water. The uneven sharing of electrons also affects how a water molecule interacts with its neighboring water molecules; the slightly negative oxygen is attracted to the slightly positive hydrogens on the next water molecule. This attraction between two different water molecules is an example of what is known as a hydrogen bond.

Hydrogen bonds are not unique to water. Many molecules can form hydrogen bonds if they have a highly electronegative atom like oxygen or nitrogen bonded directly to a hydrogen. These bonds often result in polarity, similar to what we see in water, where the slightly positive side of a molecule is attracted to a slightly negative side of another molecule. Hydrogen bonds are not particularly strong compared to covalent bonds. In fact, in a glass of water, hydrogen bonds constantly form and break between different water molecules as the liquid sloshes around. Yet this weak and fleeting

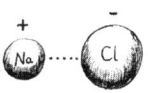

Chemical bonds can be ionic or covalent. Ionic bonds are based on the attraction between atoms with opposite charges, like the attraction between a positively charged sodium and a negatively charged chloride in table salt. Covalent bonds are formed when atoms share electrons. With a single bond, like the one found in hydrogen gas, each atom shares one electron with another atom. If two electrons from each atom are shared, then a double bond is formed, like the one found in the oxygen we breathe.

interaction has immense consequences for the behavior of water. Hydrogen bonds are the main reason that water sticks to itself, beading on a slick countertop into discrete droplets, whereas oils, which lack hydrogen bonds, spread into a very thin film that coats the surface. Hydrogen bonds also allow water to "stick" to other molecules. So-called hydrophilic (water-loving) surfaces, like that of a paper towel, attract water molecules. Hydrogen bonds also give water an unusually high boiling point. It takes a lot of heat energy to break up the hydrogen bonds between the countless number of water molecules filling a pot on a stove, so we get the saying "the watched pot never boils."

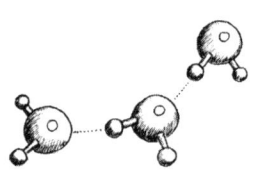

Water molecules can form hydrogen bonds (dashed lines), where the slightly positively charged hydrogen (smaller spheres) from one water molecule is attracted to the slightly negatively charged oxygen (larger spheres) from a neighboring water molecule.

Beyond their role in the behavior of water, hydrogen bonds influence the behavior of almost all known molecules, including biomolecules like proteins, fats, and even DNA. When Pauling sat in bed thinking about the structure of proteins, he found that his hand-drawn string of repeating atoms contained hydrogen atoms that might be able to make hydrogen bonds with electronegative atoms like oxygen and nitrogen. As he cut out the string of chemicals on the piece of paper, he twisted the strip into the shape of a helix, which allowed the hydrogens with their slightly positive charges to find partners that provided a slight negative charge farther up the chain. In that way, Pauling had begun to decipher one of the languages of life: the language of proteins.

THE MANY LANGUAGES OF LIFE

As late as the middle of the twentieth century, the relationship between the different components of the cell was still unclear. DNA and proteins had been too big to study in detail. At the time, scientists knew that each of our cells communicates in different chemical languages using biomolecules. For example, it was known that the language of

DNA was built on an alphabet composed of four letters: A, C, G, and T. Each of the letters is known as a "base" and has unique chemical properties. Proteins were discovered to have an even more complex alphabet composed of twenty letters, with each letter, known as an amino acid, having its own chemical properties. But even though the building blocks of the DNA and protein languages had been deciphered, how the two languages communicated and what each of them meant was still a mystery. In fact, it wasn't until 1952 that DNA was determined to be the genetic material in cells. This was the result of a ground-breaking experiment by Alfred Hershey and Martha Chase, showing that DNA from a virus is all that is needed to infect a bacterial cell and make more viruses.[6] Based on these results, Hershey and Chase concluded that DNA contained all the instructions needed to make a fully functional organism. Later in the decade, the so-called central dogma of molecular biology was established, which explained how the language of DNA is translated into the language of proteins. This solidified the view of DNA as the blueprint and instructional manual for life, and proteins as the finished functional products of these instructions.

A gene is a stretch of letters in DNA language. When translated, it becomes a protein constructed with letters of amino acids in protein language. This protein can then do its necessary work to keep the organism alive. Proteins are constructed by stringing together amino acids in a precise order. Each amino acid in the language of proteins acts a certain way when placed at a particular location within a sequence of other amino acids. But how does a language of four letters get translated into a language of twenty? As it turns out, it takes three DNA bases in a row to translate into one amino acid. For example, a DNA sequence of ATG translates into the letter M in amino acid language. Flipping the sequence of the DNA into GTA gives a different amino acid (in this case the letter V), because direction matters.

While each amino acid is unique in its structure and behavior, many have similar properties. For example, some amino acids are hydrophilic (water-loving). These amino acids prefer to interact with water and other components dissolved in water, like salts and minerals. Other amino acids are hydrophobic (water-fearing) and prefer to stay away from water. As the chain of amino acids is assembled, each amino acid resembles a bead on a flexible string that allows them to twist and bend like a rosary. Most proteins will spontaneously form a three-dimensional structure in which the hydrophilic amino acids are positioned on the outside of the protein, while the hydrophobic amino acids stay within the center to "hide" from water. This property is essential for allowing a protein to adopt the correct three-dimensional structure. A protein will carry out its function correctly only when it forms the right shape. (If a chair comes out of the factory looking like a toaster oven, it will not be a useful chair.) In this same sense, the sequence of amino acids in a row contains all the information needed for the protein to adopt its fully functional structure as it is being made by adding one amino acid at a time.

But even as different as the twenty amino acids are from each other, they all share a similar pattern. In fact, the name amino acid reflects the features they have in common. Each one contains a central carbon atom. On one side of the carbon, a nitrogen bonded to two hydrogens makes up the "amino" part, and on the other side, an "acid" part is made of a carbon with two oxygens—hence the name "amino acids."

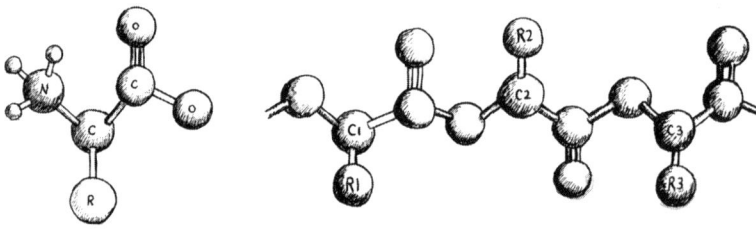

The basic structure of an amino acid consists of a central carbon atom with an amino group (made of one nitrogen and three hydrogens), an acid group (made of one carbon and two oxygens), and a side chain (shown here as a sphere with the letter R). The side chain is unique to each of the amino acids. Amino acids join together into long chains when the amino group of one amino acid connects with the acid group of another. The central carbons are labeled C1 through C3, and the side chains are labeled R1 through R3.

The difference between each amino acid lies in the third group attached to the central carbon. This is known as the side chain. Each amino acid has a special side chain composed of different combinations of carbon, hydrogen, oxygen, nitrogen, or even sulfur, giving it its unique chemical properties. When amino acids are strung together to make proteins, the amino group from one connects to the acid group of the next to form a protein that can be as small as eleven amino acids or as long as thousands of amino acids.

Amino acids in a protein are like friends joining hands in a long line. The right hand of one person joins the left hand of the next in the chain. Yet each friend is also a unique person who brings their own personality and behavior to the entire group. In addition, there are many different ways to rearrange the individuals holding hands, and each arrangement is unique. In a similar way, amino acids "join hands" by forming bonds in a specific order. The sequence in which the amino acids are arranged, one after another, gives rise to the function of the resulting proteins. This precise sequence of amino acids is a direct translation of the sequence of the DNA segment of the gene encoding that very protein.

When Linus Pauling began thinking about how proteins form the structures needed to carry out their function, it was already known how amino acids join with the amino group of the next amino acid. And as he lay in bed that afternoon in England, he began to sketch a string of amino acids connected in that very way. Pauling decided to ignore the unique side chains of the amino acids and focus on the common features. He took great care in sketching the structure as accurately as he could, using a ruler to draw the bonds and angles between each atom to scale. As the string of amino acids began to take shape on the paper, a pattern emerged. He noticed that because of the arrangement of the atoms, some hydrogens experienced an uneven sharing of electrons, leaving them with partial positive charges, much like hydrogen in water. Using his own principles of electronegativity, he postulated that the hydrogens in proteins could form hydrogen bonds just like those made by water molecules. As he cut the strips of paper containing the amino acid sequence he had sketched, he noticed that when the strip was twisted into a spiral, hydrogen bonds from one amino acid could form bonds with an amino acid farther up the chain. He called the structure the "alpha helix."[7] Like a spiral staircase, where each step is an amino acid, the helix completed one turn every 3.7 amino acids.

The discovery of the alpha helix by Pauling combined his creativity and playfulness as well as his deep understanding of molecular interactions. His ability to imagine a protein twisting to make favorable hydrogen bonds made the idea of a protein helix plausible. But at the time of the discovery, there was no evidence that such a structure actually existed. For months, the alpha helix lived only in Pauling's mind. As far as he was concerned, it was the product of a paper cutout made to pass the time while he was ill. In fact, Pauling himself was skeptical that molecules like proteins, which were much more complex than water, could "fold up" or "zip" so readily using only a pattern of hydrogen bonding. Adding to Pauling's doubt was work happening simultaneously in England, which hinted at a spiral that was like the alpha helix, but twisted differently. In Pauling's paper cutout model, it took about 3.7 amino acids to complete a full turn. But this did not agree with the British model, which proposed only whole numbers of amino acids per turn. Pauling, frustrated, put the problem away for a few months.

Even with doubt looming around his idea of the alpha helix, Pauling's interest in how proteins "fold" into their three-dimensional structures did not subside. When he returned to his lab at Caltech, he brought his ideas to his long-time research assistant, Robert Corey. Pauling, convinced that he was on to something, wanted to revisit the alpha helix. To help test his model on real-life protein samples, he enlisted the help of Herman Branson, a brilliant physicist who was at the time a professor at Howard University and was becoming a leading authority on a technique known as x-ray crystallography. Branson was invited by Pauling for a one-year fellowship at Caltech, where he would use x-rays to investigate the structure of proteins and to try to observe Pauling's alpha helix. In Pauling's lab, Branson and Corey isolated a protein from hair known as keratin and the muscle protein myosin.[8] They purified and concentrated the samples, turning them into pure crystals that they could bombard with x-rays. Then they observed how the x-rays bent as they traveled through the proteins. Using complex mathematical equations, they concluded that Pauling's proposed alpha helix was indeed present in these proteins.

In 1951, Pauling, Corey, and Branson published their results, revealing to the world the structure of the alpha helix just as Pauling had imagined it.[9] In their paper, they challenged the British model, which was later dismissed. Branson returned to Howard University and continued to investigate proteins and their structures, becoming a prominent figure in the field of protein chemistry. His pioneering work validating the structure of the alpha helix laid the groundwork for a new field of research—one that would uncover countless protein structures and lead to the development of numerous life-saving drugs. Branson continued to have an illustrious career researching sickle cell anemia, a debilitating disease that disproportionately affects those of African descent.[10] He also contributed to the success of many other Black scientists

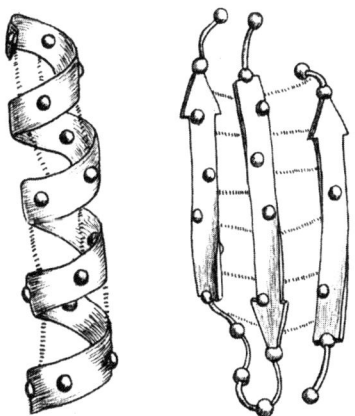

Proteins can form two main structures: alpha helices and beta sheets. Chains of amino acids can twist into an alpha helix structure that resembles a spiral staircase. They can also fold into beta sheets that resemble switchbacks. Both structures are stabilized by hydrogen bonds, shown here as dotted lines. The alpha carbons in both structures are shown as small spheres.

like himself, most prominently Mary M. Daly, who became the first woman of color in the United States to earn a PhD in chemistry.[11]

As Pauling continued to apply his ideas about hydrogen bonding to proteins, he demonstrated another element of protein structure just as important as the alpha helix. Picking up the strips of paper with the string of amino acids drawn on it, he noticed that he could lay several strips side by side in a plane. This arrangement would allow the amino acids from one strip to make hydrogen bonds with the neighboring strips in a repeating pattern. Together, the strips formed a flat sheet, which Pauling called the "beta sheet."[12] With the structures of the alpha helix and the beta sheet described and validated, Pauling and his colleagues presented two ways in which proteins can take shape and showed that these two forms can be found in naturally occurring proteins.

INTO THE FOLD

As more and more protein structures were examined, the alpha helix and the beta sheet showed up over and over again as the fundamental shapes holding the three-dimensional structures of proteins together and enabling them to function properly. In fact, the majority of the 150,000 or so protein structures known to date are composed almost entirely of alpha helices or beta sheets connected by small loops composed of short protein sequences. For example, myoglobin, a protein that stores oxygen in our muscles, is composed exclusively of alpha helices. By contrast, the antibodies produced by our immune system to fight infections are made almost entirely out of beta sheets. Other proteins, like many of the enzymes we use to digest food, contain a combination of alpha helices and beta sheets.

It may seem improbable that such incredibly complex molecules like proteins, with their vast array of sequences, sizes, and functions, could all be composed of only two simple structural elements like alpha helices and beta sheets. After all, even an average protein contains many thousands of atoms: carbons, hydrogens, oxygens, nitrogens, and sulfurs, all connected through chemical bonds. Each of the bonds can bend and twist, move and rotate, to create an unimaginable number of possible configurations. Like a very long string of beads, the atoms can theoretically take on an unlimited number of shapes and conformations as the bonds between them flex and rotate. Yet instead of flopping around like beads on a flexible string, proteins tend to adopt fixed configurations that result in either the spiral alpha helices or the flat beta sheets. The main reason for the formation of these stable structures is the hydrogen bonding that Pauling predicted. Individually, each hydrogen bond is weak, but the repeating pattern of amino acids in a sequence guarantees the presence of multiple hydrogen bonds. Together, the hydrogen bonds act like flexible scaffolding, providing a type of elastic glue that keeps the core structure stable while allowing the protein to flex and wiggle as it carries out its function.

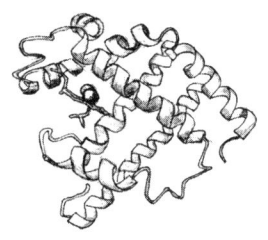

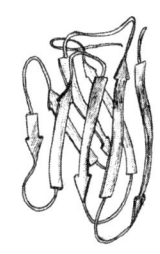

Myoglobin (*top*, PDB code 1MBO) is a major muscle protein that stores oxygen (shown as spheres). It is composed almost entirely of alpha helices. In contrast, antibodies like this one (*bottom*, PDB code 5I18) are made almost entirely out of beta sheets.

We can think of alpha helices as the beams or columns of a building, and the planes formed by beta sheets as the walls, ceiling, and floors. This combination of shapes can be combined to build a nearly unlimited number of structures while providing both strength and flexibility. For example, many structural proteins, such as those found in hair and nails, are made mostly of long alpha helices wrapped around each other like ropes, providing a rigid foundation and incredible strength.[13] Strong fibers made of long alpha helical proteins also play an essential role in protecting the integrity of animal cells. Unlike plants, algae, and bacteria, which have strong sugar-coated cell walls, our animal cells would be fragile if not for the proteins of the extracellular matrix, a strong web of proteins woven together to keep the contents of cells from oozing out. Most of our cells are wrapped in a tightly woven web made of intertwined fibers of collagen, a three-stranded helical protein. Because of its role in covering every one of our cells, collagen is the most abundant protein in the human body.[14]

Beyond their role in providing rigid structures, alpha helical proteins serve a wide array of functions in all organisms.[15] Receptors sitting on the edge of our cells are often made of alpha helices. For example, opsins, the light receptors we use to see colors, are composed of seven alpha helices arranged in a bundle. The specific arrangement of the bundle of alpha helices in an opsin and in many other types of receptors not only provides structural support; it also enables organisms to react to signals outside of the cell and to transmit those signals across the membrane and to the rest of the cell.

Another common feature of some alpha helical proteins is their ability to interact with DNA. The sticklike structure of alpha helices makes them ideal for fitting into the grooves within the twisted DNA double helix. As a result, many of the proteins that control when genes are turned on or off are made of alpha helical structures. One important example is the family of basic helix-loop-helix proteins that belong to a large family of proteins known as transcription factors.[16] Members of this family act as master regulators, determining when a gene will remain dormant and when it will be "read" and translated into a functional protein. Together, they control a vast array of genes ranging from those responsible for the circadian cycle to those involved in cell division and the progression of cancer.

Just as some of our proteins are made entirely of alpha helices, others are composed only of beta sheets. The flat surfaces of beta sheets can form barriers that separate different parts of the protein and facilitate a variety of functions.[17] An important feature of beta sheets is their flexibility. They can twist and curve as the loose hydrogen bonds holding them together flex. One common structure found in many proteins is known as a beta sandwich, where two slightly curved beta sheets sit one on top of each other like two slices of bread in a sandwich. In our bodies, such a structure can be found in a protein known as transthyretin, which

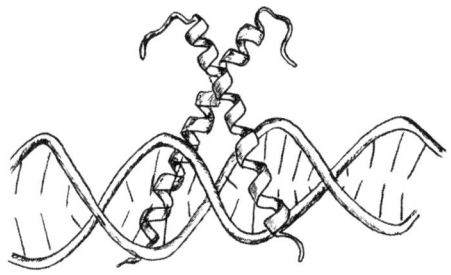

Alpha helices are common in proteins that bind to DNA. The two proteins known as Mad-Max and Myc-Max come together, each contributing one alpha helix (PDB code 1NLW). The dual proteins recognize a specific DNA sequence that influences how genes are translated into proteins.

is used to transport thyroxine, one of the main thyroid hormones, through the bloodstream. The "sandwich" structure allows the hormone to sit between the two beta sheets and be carried through the bloodstream from one part of the body to another. We also use this motif in a protein known as fibronectin, a protein that lives on the outside of our cells.[18] Fibronectin acts as a bridge from one cell to surrounding cells, and it is used to connect webs of collagen.

The flexibility of beta sheets is essential to their ability to contort into a wide variety of elaborate structures. One such example is the beta barrel, where one giant beta sheet is curved all the way around, the way a sheet of paper can be curved to make a tube.[19] Beta barrels come in many sizes. Some are small, like the beta barrel used to transport vitamin A in humans. Their small size is well suited for the long and slender structure of vitamin A to fit into the barrel during transportation. Other beta barrels are medium-sized, with perhaps the most famous example being a protein known as green fluorescent protein (GFP). The beta-barrel shape of GFP is what makes some species of jellyfish glow under UV light. Inside the barrel sits a group of amino acids that connect together to form a web that absorbs blue light and emits it back as green light. Much larger beta barrels, made of sixteen to eighteen beta strands, act as tubes for bacteria to transport sugars and other nutrients into their cells. The large hollow center of this barrel forms a channel big enough to enable the regulated flow of sugars into the cell, where it is metabolized for energy.

Combining the two structural elements of alpha helices and beta sheets has led to the evolution of the vast array of protein functions we see in all organisms. Indeed, these two basic forms can be mixed and matched to make up a nearly unlimited number of structures. Yet despite the number of organisms in the world and the immense number of proteins each organism must produce, scientists have observed only a few hundred different ways in which the two structural elements combine to make protein structures.[20] Each of these classes of structures is known as a "protein fold." It appears that the entire repertoire of protein function in the Earth's biome can be accomplished by a relatively small number of protein folds. This is because nature, in many ways, can be economical, working with the principle "if it ain't broke, don't fix it." Once a useful fold emerges, it can be used over and over, with small changes to modify the function while keeping the core structure fixed. Indeed, studies of molecular evolution have shown us that proteins with very similar structures can be

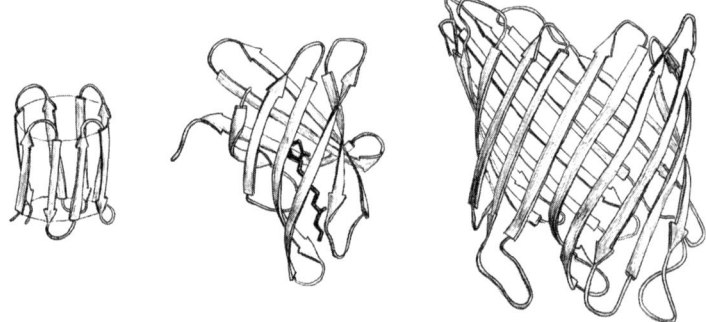

The beta barrel is a common protein structure, made when beta sheets fold into a cylinder. The beta barrels can come in different sizes; for example, the vitamin A transporter known as retinol binding protein (PDB code 1RBP) is made of eight beta strands. Vitamin A is shown inside the barrel. Large beta barrels like the bacterial outer membrane protein (PDB code 7JZ3) are made of sixteen individual beta strands. The large barrel allows the transport of molecules acting as a portal through the membrane.

modified ever so slightly to have markedly different functions, without disrupting the overall fold.

Nowhere is this principle more evident than in the so-called TIM barrel fold. This structural motif combines alpha helices and beta sheets to form a structure that is stable, yet extremely adaptable. At the center of a TIM barrel is a tube made of one curved beta sheet. The tube is reinforced by pillars of alpha helices on the outside. This single structural motif, found in all species on the planet, is thought to be ancient. Throughout the evolution of life, each organism has found different ways to tweak it for a specialized function. Even within a single organism, variations on the TIM barrel amino acid sequence result in different functions. For example, the gut bacteria *E. coli* use TIM barrels as enzymes for multiple key steps in the breakdown of sugars and for the construction of RNA and DNA.[21] In fact, most TIM barrels are used as enzymes, proteins that make chemical reactions happen at lightning speed. Like bacteria, we also use TIM barrel-shaped enzymes for many steps in the breakdown of carbohydrates from our diet. Plants use the TIM barrel for photosynthesis, and some, like the

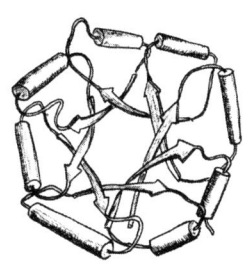

One of the most common protein folds is known as the TIM barrel. It is a symmetrical structure that combines eight core beta strands with an outer layer of eight alpha helices. The barrel structure, much like a bucket, is ideal for carrying out chemical reactions in a controlled manner.

pink morning glory, secrete a type of TIM barrel in their sap to defend against fungal and bacterial infections.[22] Some bacteria even use the same TIM barrel shape to facilitate light-emitting reactions, making them glow in the dark much like a lightning bug. Finding new ways to use the same scaffold is akin to teaching an old dog new tricks. This is a type of divergent evolution, whereby one strategy, in this case, the TIM fold, is slightly modified over millions of years, bringing about new protein functions. As these new functions appear, they usher into existence new species that can adapt to novel environments and better compete for survival.

The TIM barrel fold is an example of an evolutionary triumph, a shape that was found to be so useful that it was copied over and over, and continually modified to generate new functions—much like the Swiss Army knife, as it gained in popularity, came to offer new tools without changing its overall design. What makes the TIM barrel so efficient as a scaffold for building enzymes is its reinforced barrel, the tube that forms a cavity where reactions take place. This cavity guarantees that only the desired chemicals can enter. In some cases, too, a flap made of a beta sheet prevents the chemicals from escaping until the reaction is complete.

MOTHER OF THE RIBBON

In a lecture hall at Duke University, an unassuming figure took the stage. In a button-up shirt with a Texas tie and a long jean skirt, she began speaking to the audience of first-year biochemistry graduate students. She spoke softly and with intent, pointing to the large board behind her where giant images of proteins were projected. The lecture was like most others the students had sat through until, without warning, she unbuckled the large belt around her waist and held it up by one end. Without saying much, she began twisting the belt into a spiral. "This is an alpha helix," she said. Jane Richardson had been teaching this course for decades, and every year, she performed the belt trick. She urged her students to imagine protein chains made of sequences of amino acids as flat ribbons. The ribbons could twist into a spiral—just as she demonstrated with the belt—to form an alpha helix. Alternatively, several ribbons could lay side by side to make a flat beta sheet.

Like most biological molecules, proteins are relatively large and complex. Even an average-size protein contains hundreds of amino acids and many thousands of atoms connected to each other in different ways. If we looked at any protein structure in which the atoms were drawn as spheres connected by bonds drawn as sticks, it would look like a tangled mess with hardly any recognizable features. In the mid-twentieth century, as the structures of larger and larger proteins were being determined by advances in technology, they were becoming harder and harder to view in a way that would easily identify their basic structural features and the type of folds they adopt. Richardson's vision and creativity simplified the problem of viewing complex protein structures. She reimagined protein chains as flexible, undulating ribbons that could fold like long strips of fabric.[23]

By developing the so-called ribbon diagram, Jane Richardson revolutionized the way that scientists visualize protein structures. Using this method of visualization, it became much easier to identify protein folds just by looking at the pathways the ribbons took. The ribbon diagram is today the most common way to show protein structures in scientific articles and books (including this one). The power of the ribbon diagram lies in how it elegantly, yet accurately, represents the topology and fold of any protein. In the ribbon diagram, alpha helices resemble twisted strips of fabric, while beta sheets are constructed using flexible, flat arrows lying side by side. The direction of the arrows indicates how that protein chain moves from the first amino acid to the last.

As a child, Jane was always interested in science. She built her own telescope and followed celestial objects as they rotated in the sky. At seventeen, she observed the Russian satellite *Sputnik* and made calculations of its orbit, which earned her a third-place prize in the prestigious national Westinghouse Science Talent Search. In college, she opted to study philosophy, receiving a master's degree from Harvard University. In the 1960s, Jane began working at MIT with her husband, David Richardson, who was studying protein structures. At the time, there were only a few known protein structures, and Jane began drawing all of them by hand, representing the chains as ribbons traveling through each structure.

Jane's ribbon drawings of proteins were well received by the scientific community. From a mess of atoms and bonds, a protein's fold could be easily recognizable. One of the most well-known examples is her hand drawing of the TIM barrel. Jane's creativity transformed a complex jumble of thousands of atoms resembling a big fuzzball into an elegant structure with an order and flow to it. Looking at the TIM barrel, one can only be astonished by how such a complex molecule can fold so neatly, with exquisite symmetry. From her drawing, it became clear how the shape of the protein, with its characteristic barrel at the center, forms a tiny reaction vessel where metabolic processes occur. The iconic structure is as much a work of art as it is a product of scientific discovery. On November 19, 2009, Jane's TIM barrel was featured as the Wikipedia picture of the day.

Yet Jane's ribbon diagrams are not merely ways to represent the structures of complex molecules as visually manageable images. Over the decades, Jane and Dave Richardson have developed various methods for protein structure visualization as a way to examine the natural patterns that protein structures are most likely to adopt. Their work has also revealed structures that are not likely to occur. This has indeed been extremely useful in refining the results from experiments on protein structures and guiding the discovery of new structures. It has also been crucial for identifying and "repairing" mistakes in reported structures.[24]

COLORING OUTSIDE THE LINES
While the ribbon diagram offers both an elegant and simple means of showing the contours of protein structures, there are many other ways to describe the structures of a given protein visually, each of which provides a different level of detail and complexity. One of the simplest is to draw the center of each amino acid as a sphere connected to the next amino acid by a single line. The advantage of this representation is its simplicity; a helix can be easily identified by the way the amino acid spheres form a spiral loop. But this bare-bones representation does not take into account the identity of the amino acid at any given position, which means that a large amount of information is left out.

If we go the other direction, toward a more nuanced and information-rich depiction, we can add to the complexity of the basic skeleton of the alpha helix in the ribbon diagram by adding the amino acids, with side chains that point out horizontally, perpendicular to the axis of the helix. Some amino acids will have small side chains; others will be large and bulky. Some will resemble zigzagging wires as the atoms in their side chains extend outward; others will form five- or six-membered rings. In the simplest form of the ribbon diagram, an alpha helix from one protein looks almost identical to an alpha helix from another protein. But once the side chains are added, differences begin to appear. Pardaxin, the small protein we saw in Chapter 1 that is secreted

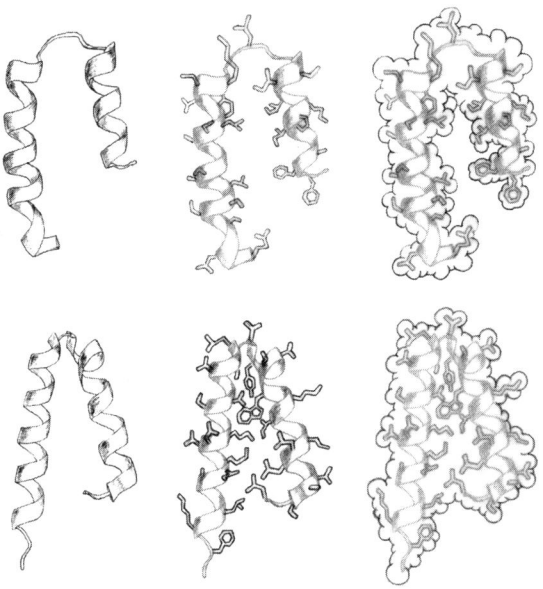

There are several different ways to show protein structure. The ribbon diagram shows how the backbone of the protein twists into its basic structural element. Shown here are two proteins, pardaxin from the Moses sole (*top row;* PDB code 2KNS), and YscF from the bacteria that causes the bubonic plague (*bottom row;* PDB code 2P58). In the ribbon representation, both proteins look similar, composed of two helices connected by a short loop. A second representation is made by adding the amino acid side chains, represented here as "sticks and branches." Once amino acid side chains are added, differences begin to emerge. A third representation is made by adding the surface volume of each of the atoms in the proteins. This is like shrink-wrapping each of them, and in this representation still more differences can be seen—the two helices of pardaxin are separate from each other, whereas the two helices of YscF merge together through the interactions of the side chains.

by the Moses sole to repel sharks, is made of two short helices connected by one loop. This simple helix-loop-helix structure is not unique to pardaxins. It is found in many other proteins, some with vastly different functions. For example, YscF is a protein found in *Yersinia pestis,* the bacterium that causes the bubonic plague.[25] This protein is part of an assembly of proteins that come together to form a needle, which the bacteria uses to inject toxins into the cells of their hosts. YscF also has the characteristic helix-loop-helix structure of pardaxin. Viewed side by side using the ribbon diagram, YscF and pardaxin look almost identical. Yet one comes from bacteria, and the other from a fish, and they each function in very different ways.

While the basic structural features of the two proteins are nearly identical, what makes the two proteins function differently is the amino acid composition of each protein. One way to see this is by adding the structures of the amino acid side chains on the backbone "skeleton" ribbon diagram. When we do so, we see the structural differences between YscF and pardaxin. YscF is made of amino acids that have larger side chains than those of pardaxin. The amino acid side chains of YscF also appear closer to each other, bringing the two helices together, while the shorter amino acids of pardaxin leave a gap between its two helices.

The differences between the two proteins become even more apparent when we represent every atom in the side chain as a sphere. This representation is akin to "shrink-wrapping" the protein to show its real volume. It is called a space-filling representation because it represents the volume of each atom as a sphere. While the space-filling representation is a more accurate way of depicting the true volume of a protein in space, it makes it difficult to see how the individual atoms are connected or to even make out the way the helix winds as you would see in a ribbon diagram. When we depict pardaxin and YscF in the space-filling diagram, we see that pardaxin forms an inverted U resembling a horseshoe, whereas YscF, with nearly no gaps between the two helices, looks like one contiguous cylinder.

Looking at the proteins even closer, we can begin to add colors to identify each sphere by element—carbons in one color, nitrogens in another, then oxygens, and so on. This representation allows us to see that the structures are not the only place where variations can be seen; there are also differences in the type of atoms that make up the amino acid side chains of each protein. YscF has more oxygens and nitrogens on its surface. This makes YscF more hydrophilic. It dissolves readily in water, which makes up most of the fluid surrounding its host cell. Pardaxin on the other hand, with fewer oxygens and nitrogens, is more hydrophobic. This property enables it to seek out the oily (lipid) membranes of a shark's cells, poking holes in them and rendering the shark temporarily paralyzed.

The chemical properties of the side chains on the two proteins are the key to how each protein functions, giving an organism a crucial survival advantage in its ecological niche. The ribbon diagram can be viewed as the backbone, or skeleton, of the protein. It defines the overall fold, while the side chains represent the appendages, the facial features, and protrusions that wiggle around as the protein carries out its specific function. The shared backbone with different side-chain "appendages" is a common motif in protein biology and an example of divergent evolution, in which a stable fold is used as a scaffold for building newer functions, by modifying the amino acids while keeping the overall framework.

Using a limited number of universal folds, variations on the amino acids that decorate the surfaces of proteins have produced an immense number of different functions. Through the millennia, small changes that allow for new functions while preserving the stability of the original scaffolds have been favored, enabling organisms to proliferate and thrive in their unique environments. Our own evolution and the differences we see among ourselves and all other forms of life on Earth are the result of incremental changes in the sequence and composition of proteins, the building blocks of every living cell.

UNCOOKING AN EGG

Even though we cannot see proteins and the structures that they make with our naked eye, we encounter them daily. After all, many of the foods we consume, such as beans, milk, and animal meat, are rich in protein. When we eat them, we break down the proteins into amino acids to build up our own proteins. Every time we crack an egg, we encounter proteins. The thick, clear goop of egg white is a concentrated solution of dissolved proteins. These proteins serve as a protective layer for the bird's embryo and provide food for its growth. Before an egg is cooked the egg white is clear, since all its proteins are completely dissolved. Each protein is folded into its particular 3D structure made of alpha helices or beta sheets, and each one is swimming in a solution of salty water. The overall structure of each protein guarantees that the water-fearing hydrophobic amino acids are protected from the surrounding water molecules, while the hydrophilic, water-loving amino acids point outward, facing the water. But when an egg is cooked, a very familiar transformation takes place. The clear solution of the egg white becomes semi-solid, opaque, white—and delicious.

What happens to the proteins of the egg whites when they are heated is a process called denaturation. This occurs when the heat disrupts the natural structure of the proteins by destroying the arrangement of the alpha helices and beta sheets. As more and more heat is applied, the hydrophobic insides become exposed to the surface, and the proteins are in a sense turned inside out. When the exposed hydrophobic regions of the denatured proteins seek refuge from the surrounding water, they begin to stick together and aggregate, scattering light and producing the cooked egg appearance. By this point these cooked proteins have lost not only their natural structure, but also their function. Heating any protein will eventually denature it. The energy from the heat breaks up hydrogen bonds, including those that hold the alpha helices and beta sheets together. Our own proteins will also denature if enough heat is applied. Our bodies function best at 37°C (98.6°F), the temperature at which our proteins are fully folded and functional. We have a built-in temperature control system as well as a natural alarm—heat receptors on our skin—to warn us about dangerously hot objects and to protect our proteins from unfolding.

When the alpha helix and the beta sheet were first discovered, it quickly became clear how heat can disrupt their neatly arranged structures. The fragility of life became evident; an increase of even just five to ten degrees would mean peril for our most important molecular machines. No one at the time imagined that life could exist at high temperatures. That continued to be the common understanding until the late 1960s, when Thomas Brock, a professor of microbiology at Indiana University, made a shocking discovery. While working at Yellowstone National Park, he and his undergraduate student Hudson Freeze discovered a species of bacteria that thrives inside the Mushroom Pool, a geyser basin where water temperatures reach a scalding 80°C (176°F). The two scientists named the new bacteria *Thermus aquaticus*.[26] The discovery of such a bizarre new species sent shockwaves throughout the scientific community. It showed for the first time that life can thrive at extreme temperatures. Soon after, several other species of bacteria were discovered near volcanoes and hot springs. Others were found in boiling lakes of acid.[27] These bacteria were given the name *thermophiles*, meaning heat-lovers, to distinguish them from *mesophiles*— organisms like us that just can't take the heat.

As more thermophiles were discovered, scientists wondered what the proteins inside those extreme organisms looked like. How could they withstand extreme heat that would literally cook an egg and render most proteins a denatured mess? At first, it was thought that the proteins from thermophiles had their own shapes, which differed from the alpha helices and beta sheets that mesophilic proteins take on. But as the structures of thermophilic proteins were investigated, they were discovered to have the exact same types of alpha helices and beta sheets found in all other known proteins.[28] And just like the mesophilic proteins, the helices and sheets were held together by hydrogen bonds. With no unique features distinguishing the structures of thermophilic proteins from their mesophilic cousins, the reason that one set was far more stable at high temperatures remained elusive.

In 2001, a group of scientists discovered a new species of thermophilic bacteria in a hot spring near Tengchong in China, a region known for its active volcanoes. The scientists named the new species *Thermoanaerobacter tengcongensis* or *Tte* and determined that its optimal growth happens at around 75°C (167°F).[29] A few years later, a group from Duke University isolated one of the proteins from *Tte,* a protein used to grab onto a sugar known as ribose. The scientists wanted to understand why proteins from this organism were resistant to denaturation at high temperatures. Once they determined the structure of the ribose binding protein from *Tte,* they compared it to the structure of the same ribose binding protein in *E. coli,* the mesophilic bacteria found in our gut. Even though the two proteins had different sequences, they had a nearly identical structure. When the scientists took a closer look, however, they found that while the amino acids near the core of both proteins have the exact same sequence, farther out toward the surface of the proteins there were differences in the amino acids. This suggested that the thermophilic protein was interacting with water in a way that provided stability at high temperatures. Indeed, the amino acids on its surface formed a strong network of ionic bonds, connections between positively and negatively charged amino acids. These bonds acted like a protective web of connections, preventing the heat from unfolding the protein. The *E. coli* protein, which had fewer ionic bonds, had a weaker network and so was more susceptible to unfolding.

This work was not the first attempt to understand what makes thermophilic proteins more stable. Other studies had found that expanding the network of bonding between the amino acid side chains greatly enhances protein stability.[30] What made this study so interesting is that it showed how two proteins with nearly identical structures could have such different stability due to their ionic bonding networks. And as it turns out, the ionic interactions between positively and negatively charged amino acids are just one way to stabilize proteins. Other studies of the stability of mesophilic and thermophilic proteins have found differences at the core of the proteins' 3D structures. Many thermophilic proteins fold into tightly packed structures due to what are known as "hydrophobic interactions." At the core of these super-stable proteins, large patches of hydrophobic amino acid side chains come together.[31] This prevents the protein from unfolding, which would expose these hydrophobic patches to the water surrounding the protein. The forces that "hide" these hydrophobic amino acids from water can keep a protein from denaturing, even at high temperatures. And so, ionic interactions at the surface and hydrophobic interactions at the core of the protein can combine to offer different ways for thermophilic proteins to maintain their shapes at high temperatures. The contribution of each of these factors to protein stability varies from one protein to another.[32]

Understanding how thermophilic proteins survive at high temperatures may be key to understanding how life emerged on our planet. Around 4.5 billion years ago, the conditions on Earth were harsh, with frequent volcanic activity, high temperatures, and a lack of oxygen. Early life had to thrive under extreme conditions, especially the high temperatures that would denature most of our own proteins. Such conditions are still found in some pockets of our planet. Most exist near volcanos or hot springs at the bottom of the ocean. The Mariana Trench is a deep gash that runs along the floor of the Pacific Ocean, between Japan and Indonesia. It is where the Pacific and the Mariana tectonic plates meet and where large cracks in the Earth's crust allow massive hydrothermal vents to spew geothermally heated water, various minerals, and occasionally hot lava into the ocean. Clouds of toxic sulfur and methane gases billow out of the cracks into the water at temperatures reaching nearly 400°C (752°F). At these extreme depths, where even Mount Everest would be completely submerged with more than a mile to spare, the water pressure is immense.[33] Yet life thrives even in these unlikely conditions.

In fact, a mind-blowing diversity of species congregates around the vents of the Mariana Trench.[34] For these species, the vents are a life-giving fountain that produces one of the most unusual (to us) ecosystems on Earth. Thermophilic bacteria use proteins to feed on the toxic gases and other nutrients. As they thrive, the bacteria support an entire community of organisms, feeding tube worms, shrimp, and clams, which in turn are consumed by larger predators like fish and crabs.[35] Completely isolated from the rest of the world, these bacteria are unique in many ways. Not only do they thrive in extreme temperatures and pressures, but they also do not get their energy from sunlight like bacteria in other ecosystems; instead they acquire it from the chemical energy trapped inside the Earth. As they feed on Earth's nutrients, thermophilic proteins drive the chemical reactions necessary for their survival. Such conditions, while found only in small pockets today, were likely very common during our planet's tumultuous beginnings. Learning about how these molecular machines manage to thrive under such immense pressures and high temperatures, then, can not only help us understand the resilience of life, but also provide clues to how life itself emerged on our planet.

Beyond questions about the origins of life, thermophilic proteins have played an important role in the advancement of biotechnology. The discovery of *Thermus aquaticus* by Thomas Brock opened the door for proteins to be used in ways that had previously seemed impossible. Proteins from thermophiles could now function under conditions where heat is required. The isolation of one of the proteins produced by *Thermus aquaticus*, Taq DNA polymerase, came to revolutionize the field of molecular biology. It streamlined a process known as PCR, or polymerase chain reaction, in which billions of copies of a piece of DNA can be made within a couple of hours.[36] Before DNA can be copied, it must be unraveled by breaking the hydrogen bonds between its two strands. PCR requires repeated heat treatments. It turned out that the DNA polymerase protein from *Thermus aquaticus* was ideal for the task. DNA polymerase is an enzyme found in all living organisms and is responsible for copying DNA during cell division. What's unique about the *Thermus aquaticus* DNA polymerase is that it can tolerate high temperatures without unfolding. The protein made PCR easier to use and revolutionized the blossoming field of molecular biology in the 1980s and 1990s. The ability to easily make copies of DNA ushered in a new era of "molecular cloning," which gave scientists the power to study how genetic information is translated into functional proteins and how mutations can lead to disease. *Thermus aquaticus* DNA polymerase was also essential for sequencing the human genome and the genomes of many other organisms, helping us to decipher the blueprint for life.[37] PCR continues to be one of the most widely used techniques in molecular biology; it is used in medical research, diagnostics, forensic investigations, and even the detection of COVID infections. Other thermophilic proteins are used in industrial processes that require high heat conditions, such as making drugs, paper, and textiles. Today, we can even find thermophilic proteins in most household detergents, which allows them to clean well even when used with hot water.[38]

For most proteins, denaturation is irreversible. Unable to adopt their original structures, denatured proteins are essentially dead. A cooked egg is a perfect example. Once the proteins aggregate into the opaque white substance, there's no turning back: it's impossible to un-cook an egg. Or it least it was until 2015, when a group of scientists from Australia and the University of California, Irvine set out to do just that.[39] Their task was to take the aggregated proteins in a cooked egg and bring them back to their original structures. Starting with a boiled egg, the group took the white solid mass of aggregated proteins and dissolved it in a solution of urea, a chemical that can break apart the protein aggregate. But this was only half the task. Even though the proteins were no longer aggregated in a clump, the individual proteins had not returned to their original 3D structure, meaning that they still could not function properly. So the researchers put the proteins under shear stress: they passed them through microscopic channels, and along the way squeezed the proteins between two surfaces in a manner similar to how we press against our food as we chew. When applied in a controlled manner, this force was able to refold the unfolded proteins and to restore their structure as well as their function.

While uncooking an egg may seem like a frivolous exercise, the approach could have wide-ranging implications. For example, when producing protein-based drugs like insulin, oftentimes a significant portion of the proteins undergoes denaturation. Understanding how proteins fold, how they denature, and how to bring them back to their functional form will help improve the production process for many proteins used as therapeutics.

Investigating how such complex molecules like proteins take shape requires not only knowledge of the rules of chemistry and physics, geometry, and biology, but also a fertile imagination, rich creativity, and unstoppable curiosity. Early research into proteins revealed their vast complexity. While it took a lot of research to probe the structure of proteins, it took an equal amount of ingenuity to describe those visually in ways that can help us understand how they work. Jane Richardson's contributions to our understanding of protein structure, especially her invention of the ribbon diagram, is a prime example of how a deeply creative mind, grounded in the arts, can provide insights into some of science's biggest questions. She was able to combine art and science to present a picture of proteins that is now used by most scientists to uncover their structures.

Artists and scientists are often viewed as two very different types of people who lie on opposite sides of the creativity spectrum. Some may think of scientists as rigid and constrained by facts, unable to access attributes of creativity and imagination that seem to belong exclusively to artists. But nothing could be further from the truth. Both scientists and artists share a lot in terms of thought processes, reliance on creativity, and ability to innovate. Research shows that most successful scientists are endowed with an artistic ability or a creative talent.[40] Other research shows that doctors, when trained in the arts, are better at making diagnoses.[41] There is no doubt that art is an essential component of science, and that the two have always worked hand-in-hand to produce the most innovative solutions to problems of all kinds.

Over millennia, scientists have demonstrated that a creative mind is one that can find form in chaos. And nowhere is this demonstrated more than in the discovery and illustration of protein structures. It was Jane Richardson's creative mind that saw through the mess of atoms that formed the bulk of proteins and developed an elegant, informative way to represent their beauty. It was the imagination of Linus Pauling, who, while sick in bed, brought the shape of a helix into existence out of a set of chemistry rules—and then, with Robert Corey and Herman

Branson, set the stage for the field of structural biology, which transformed our understanding of one of the most fundamental languages of life: the language of proteins. And it was the curiosity of Thomas Brock that led him to find life in the most unlikely places, which in turn drove the discovery of thermophiles, the streamlining of PCR, and the molecular biology revolution of the late twentieth century. Collectively, the contributions of these scientists have allowed us to look deep within our own cells for answers to the ultimate question of who we really are.

5

LIVING
SHAHIR S. RIZK AND MAGGIE M. FINK

IN A SMALL CLASSROOM in the basement of
Northside Hall, on the campus of Indiana University South Bend, an
elegant woman spoke to a group of college students. The bright fluorescent lights in the ceiling highlighted her snow-white hair as she spoke
with a commanding voice. Gretchen Anderson talked about her favorite
bacteria, an obscure species of soil dwellers named *Alcaligenes faecalis*.
The students squirmed when they realized that the bacteria were named
after where they were found—sheep feces. Gretchen smiled and continued talking about this strange creature. Not many people have heard
of this species of bacteria. What makes them unique is where they live.
Alcaligenes faecalis are found in toxic environments where most living
things would perish. The bacteria, it seems, can tolerate extremely high
levels of arsenic; in fact, they seemed to thrive on the toxic metal.[1]
Gretchen explained how arsenic was the most commonly used pesti-
cide in the early parts of the twentieth century. It was sprayed on corn,
rice, and apple orchards by the ton to control the aphids that plagued
these crops, a practice that continued in many parts of the United States
well into the 1980s.[2]

While arsenic was effective in killing pests, it posed serious health
effects for those working in the fields and for those who consumed
sprayed crops. It also left many fields contaminated for decades, and
others nearly devoid of life. Except for a few extreme creatures like

Gretchen's favorite bacteria. In her small lab, Gretchen studied how the bacteria thrive in such toxic environments. Her work revealed that instead of avoiding arsenic, the bacteria have a special attraction to it. They mop it up from the environment by latching onto it, then transform it into a less toxic form. Gretchen showed that this transformation was the result of a special protein—an enzyme called arsenite oxidase that is unlike any we have in our bodies but is key to the bacteria's survival. Every summer, two or three college students spend weeks in her lab studying how arsenite oxidase works on the molecular level. The hope is that one day the tiny enzyme will help restore fields that had been sprayed for decades with a deadly poison.

Every organism that has ever existed, from the smallest bacteria to the giant sequoia, makes a special class of proteins that we call enzymes. These specialized proteins are biology's tiniest machines. Enzymes make chemical reactions happen by either breaking or making bonds. They are the engines of life and the bio-agents of change, continually constructing and deconstructing almost all components of life needed for survival. Some enzymes can be used for demolition to break down food, digest meals, neutralize toxins, or protect from chemical and biological threats. Other enzymes are master builders, used to construct biomolecules by assembling smaller components. Just as arsenite oxidase helps bacteria survive in a contaminated field, enzymes in our bodies help us live, grow, move, and reproduce. When a cell divides, enzymes string together A's, C's, G's, and T's to make two identical copies of DNA. Our heart muscles require the action of enzymes to channel chemical energy into a beating rhythm, and our stomachs rely on countless enzymes to digest carbohydrates, proteins, and fats in our food. The kidneys filter salts and toxins into urine through the action of enzymes, and fat cells build and store lipids for later use with the help of a precise set of specialized enzymes.

All enzymes are proteins made of the same building blocks—amino acids—strung together in a unique order, then folded into a precise shape to perform a certain function. What makes enzymes special among proteins is their ability to carry out chemical reactions. The protein hemoglobin, for instance, which carries oxygen from our lungs to our tissues, is an example of a protein that is not an enzyme. It binds to oxygen but has no way to change it into a different molecule. It simply releases oxygen, unchanged, where it is needed in the tissue. Enzymes, by contrast, bind to a molecule and change its chemical composition before releasing it as a new molecule.

Enzymes, like all catalysts, oversee a transformation without themselves being changed in the process. For example, the sugar maltose, found in starchy foods and beer, is made of two glucose sugars connected through a chemical bond. In our stomachs, the enzyme maltase (many enzymes end with the suffix "-ase") breaks the chemical bond in maltose and separates the two glucose sugars from each other. Once the two glucose molecules are separated, each can be digested to produce energy. This may seem like a simple process—break one bond between two sugars. And in fact, the process is spontaneous: given enough time, the reaction will occur. But saying a process is spontaneous does not mean that it will happen quickly. While the breakdown of a sugar like maltose is spontaneous, it is estimated to require about 20 million years if left alone.[3] But add a pinch of maltase, and the reaction happens in less than one hundredth of a second—more than 70 quadrillion times faster.[4]

Without this enzyme acting as a catalyst, it would be impossible for us to capture any of the energy in the maltose sugar, since breaking it down would take about three hundred thousand lifetimes. Like maltase, all enzymes speed up the rate of chemical reactions immensely, and it is this incredible ability that makes life possible. Even more amazing is that enzymes are unchanged in the process. After breaking down one maltose, the very same enzyme can go on to break down another and another and another. In fact, one maltase enzyme can break down about 140 maltose molecules before you can say "one Mississippi."[5]

Enzyme "catalysis" is the process by which enzymes grab onto specific molecules and convert them into other molecules. The starting molecule is known as the substrate, and what's released is known as the product. In the reaction carried out by maltase, the sugar maltose is the substrate, and the two glucose sugars that result from breaking the bond are the products.[6] In every organism, enzymes work together as a team to accomplish a specific mission for which every enzyme has a predefined job. The product of one enzyme is often the substrate for the next in a long sequence of chemical modifications. The reaction carried out by maltase, for example, is the first in a stepwise process of digesting many of the sugars we eat. The two glucose molecules produced by the breakdown of maltose are each used by the next enzyme in the chain. At each step, the glucose is further modified by additional enzymes until energy is completely extracted from the sugar: the leftovers are then discarded as water and the carbon dioxide we exhale.

A chain of enzymes, like that used to break down sugars, resembles workers at a conveyor belt on a factory floor. In living systems, this is known as an enzymatic pathway—a series of consecutive chemical reactions, each one catalyzed by an enzyme. Some pathways, like those we use to digest the different components in our food, are dedicated to deconstruction or disassembly. Other pathways are more like assembly lines used to build biomolecules. For example, after a big meal, short chains of carbon scavenged from our food are linked together to make lipids (fats), which we store to produce energy or use to insulate us from the cold. Other bio-assembly pathways lead to the production of hormones, steroids, and even DNA. Nearly every biomolecule on our planet is made by the action of enzymes. Over billions of years, each enzyme in its pathway has become adapted to carry out a specific reaction, and this evolution is still taking place, with enzymes continuing to become more specialized and attuned to their jobs of assembly or disassembly.

THE MOTHER OF ENZYMES

In 1912, less than one year after the sinking of the *Titanic,* a young woman boarded a transatlantic ship bound for Germany. Unaccompanied by family or friends, Maud Menten was on a mission; she was headed to Berlin University. By then, she was already one of the first

Canadian women to have earned a medical degree. At the time, women were not allowed to work in research labs in Canada, so she made the trip to Berlin University to work with Leonor Michaelis, a scientist who was studying the nature of enzymes. When she arrived, Menten began examining the behavior of these mysterious biological entities. Most of what was known about enzymes at the time was based on early experiments by Louis Pasteur, who had shown in the mid-1800s that substances in yeast cells could catalyze fermentation due to what he called a mysterious "vital force."[7] In 1877, a German physiologist named Wilhelm Kühne coined the term enzyme, which he derived from the Greek word meaning "in yeast."[8] Yet no one knew how enzymes worked or what they were made of.

Michaelis and Menten began by measuring how enzymes accelerated chemical reactions. They viewed enzymes as biological machines that resembled the engine of a car. Just as when a car's engine receives more fuel, it moves faster, they noticed that as more and more substrate was added to the enzyme, the reaction began to speed up. They also observed that, like cars, all enzymes have a top speed, or maximum rate; no matter how much fuel is added, the engine will not go any faster. Michaelis and Menten called this "the maximum speed" and notated it as "V_{max}."[9] Moreover, different enzymes had different speed limits. Some were like Ferraris; others were more like Volkswagens.

Over the course of one short year, Michaelis and Menten worked relentlessly to elucidate the nature and behavior of enzymes. In addition to showing that an enzyme has a maximum speed, they found that every enzyme requires a specific concentration of substrate to work at half of its V_{max}. This substrate concentration is constant for each enzyme, and they called it "K_M." By measuring V_{max} and K_M for an enzyme, scientists can learn about the top speed for an enzyme and how much substrate it takes to get there. In 1913, Michaelis and Menten published a mathematical equation that summarized the relationship between V_{max} and K_M. The equation enabled scientists to study and classify an immense number of enzymes, paving the way for the discovery of innumerable drugs and medicines. Today, knowing the V_{max} and K_M

values for enzymes helps researchers to decipher entire pathways for biosynthesis and identify the key steps in the long assembly lines of life's most important molecules. This knowledge makes it possible to develop new drugs that can inhibit not only processes like cell growth in cancers, but also the production of molecules responsible for inflammation. Enzymes orchestrate nearly every reaction in every living organism. Understanding how enzymes work, how fast they carry out reactions, and how they connect in long metabolic chains provides a picture of the dynamic inner workings of life itself.

After her brief stay in Berlin, Menten made the trip back across the Atlantic, this time to study biochemistry at the University of Chicago. After completing her PhD there in 1916, she struggled to find a research position in her home country of Canada, so she returned to the United States and joined the University of Pittsburgh's Children's Hospital as a clinical pathologist. During her time there, Menten worked tirelessly treating children, giving lectures, teaching lab courses, and developing a world-class research program. She applied her work on enzymes to study the properties of DNA, develop cancer treatments, and understand the behavior of hormones, among other pursuits. She published over a hundred articles and rarely worked fewer than eighteen hours a day. She was also interested in astronomy, spoke multiple languages, and her artwork decorated the hallways of the university. Yet for all her hard work, she received little recognition. In fact, Menten was not promoted to the rank of full professor until thirty years after joining the university, just two years before she retired.

Menten's contribution to our understanding of enzymes was truly groundbreaking. It's astonishing to realize that she, along with Michaelis, found a way to study enzymes before anyone even knew what enzymes were made of. In fact, it wasn't until 1930 that enzymes were shown to be made of proteins. Menten's pioneering work predated that discovery by seventeen years and Pauling's alpha helix structure by nearly four decades. Even with all the advances in protein science and the study of enzymes, the Michaelis-Menten equation is taught in every biochemistry course more than a century after it was proposed.

ENGINES OF LIFE

When a cell divides, thousands of enzymes come together to coordinate the process. At the heart of this process is an enzyme in charge of duplicating DNA, ensuring that two identical copies are made, one for each of the two new daughter cells. This enzyme, known as DNA polymerase, has a big responsibility.[10] It must copy a long string of letters known as DNA bases with high fidelity. Mistakes could lead to disease-causing mutations, which could be passed on to the next generation. The enzyme must also work extremely fast. Packed into each of our cells is a DNA string three billion bases long. If stretched out, the DNA in any one of our cells would be about six feet long. If all the DNA in your cells was linked together end to end, it would be seven hundred times longer than the distance to the sun. Luckily, we don't have to copy the DNA of all our cells at the same time. But to copy the DNA of a single cell, we still need to replicate three billion bases in a row, and it has to be done quickly and with the fewest mistakes possible.

This might sound daunting, but DNA polymerase is more than up for the challenge. A single DNA polymerase enzyme can copy about a thousand DNA bases in less than a second. With multiple DNA polymerase enzymes working together in synchrony, the entire DNA in the cell is copied in less than a minute, rendering the cell ready to divide. Even at such a dizzying speed, DNA polymerase is extremely accurate in copying the DNA. It makes a mistake only once every billion bases it copies. Imagine copying the entire text of roughly ten thousand books by hand before making a single letter typo. The incredible speed and accuracy with which DNA polymerase works makes it one of the most efficient machines on the planet. In fact, we use the remarkable properties of DNA polymerase in many aspects of biotechnology. The PCR test used to detect COVID-19 utilizes DNA polymerase to make many copies of the genetic material of the virus so that it can be detected in a sample. We also use DNA polymerase in determining paternity and in a wide range of forensic tests to identify a perpetrator by the presence of their DNA at the crime scene.

Studying how enzymes work can give us a glimpse into their personalities. It can also help us develop tools for developing new medicines. In the early 1980s, when a mysterious sickness appeared among young men, depriving them of their immunity, medical researchers began working immediately on identifying the basis of the disease. Soon the illness became known as acquired immune deficiency syndrome (AIDS) and was found to be caused by the human immunodeficiency virus (HIV), a pathogen that infects the very immune cells dedicated to fighting infections. This leads patients to become vulnerable to all sorts of infections from bacteria, viruses, and even fungi that would normally be no match for a healthy immune system. The HIV virus is tiny, about a hundred times smaller than the diameter of the immune cells it infects.[11] It can carry the genes for only fifteen proteins of its own.[12] Like all viruses, it relies on the enzymes of the cell it infects to make more of itself, even by hijacking the patient's own DNA polymerase. The few proteins unique to the virus are the ones responsible for its assembly into rounded viral particles. They make up the outer layer of the virus and the hooklike proteins it uses to latch onto immune cells to infect them. Most of the virus's proteins are not made one at a time. Instead, they are produced as one long string of proteins, attached end to end. For these proteins to function properly, they must be separated from each other, and to do this, the virus has a unique enzyme, called HIV protease. This enzyme is the master regulator of the entire biology of the virus. It breaks apart the string of proteins into individual functional units, setting off the life cycle of the virus as each protein begins to do its designated job. The action of HIV protease is vital to the progression of HIV. Without it, the virus would not be able to propagate inside a patient's body.

Structural analysis of the HIV protease enzyme revealed two identical protein subunits made of alpha helices and beta sheets.[13] The two subunits are locked together in a handshake. With its two identical subunits, the enzyme latches onto the long string of viral proteins and cuts them at precise locations into individual functional units. It became clear that because HIV protease activates all of the

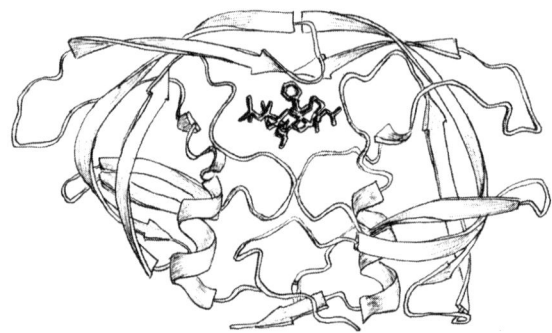

Many HIV drugs are potent inhibitors of the viral enzyme known as the HIV protease. The function of this enzyme is crucial to how the virus survives in a patient's body. The drug known as saquinavir (shown in dark shading) binds at the center of the HIV protease where the substrate would normally bind and blocks its activity (PDB code 4QGI).

virus's proteins, then stopping its activity would surely slow down the infection. Scientists began to design molecules with structures that would fit inside the protease with the hope of blocking its ability to process the viral proteins. In 1987, a molecule named saquinavir was tested on AIDS patients and showed promising results. The drug was approved in 1995, giving hope to thousands suffering from a disease that had been considered a death sentence. That same year, many more HIV protease inhibitors became available to AIDS patients. Even today, many of these drugs continue to save countless lives and improve the quality of life for AIDS patients.

The strategy of blocking enzymes to stop diseases is not limited to treating HIV. In fact, enzyme inhibitors are used to treat countless diseases and conditions ranging from cancers and autoimmune diseases to bacterial and fungal infections.[14] Enzyme inhibitors are also used to treat inflammation, allergic reactions, heart disease, and respiratory illnesses. In all these cases, the strategy is the same: a small-molecule drug blocks the function of an enzyme. But long before humans designed drugs to stop enzymes from working, microorganisms developed countless inhibitors of their own. For billions of years, these species have crafted inhibitor molecules that target the enzymes of rival organisms to gain a survival advantage.[15]

This constant competition between microorganisms was key to the discovery of one of the earliest and most effective antibiotics—penicillin. In 1928, after returning to his lab from a vacation, the Scottish microbiologist Alexander Fleming noticed that one of the Petri dishes he had used to grow bacteria was contaminated with a mold called *Penicillium rubens.* Instead of throwing the dish away, Fleming took a closer look at how the two organisms had grown on the dish. He noticed that the growth of the mold had pushed away the bacteria he was growing, *Staphylococcus aureus,* a type found in a wide range of illnesses, from minor skin infections to bacteremia and sepsis. Fleming wondered if some chemical the mold was producing had inhibited the growth of the bacteria. Fleming's discovery was not an immediate breakthrough; instead, it was the beginning of a journey to harness the power of penicillin. He published his findings in 1929, but it would be many years before the actual chemical responsible for killing bacteria was identified, and even longer before it could be purified and used in isolation as a drug.[16] This was common in research of the early twentieth century. An organism or parts of it were seen to have some type of chemical activity—in this case, antibiotic activity—but the tools to isolate the exact molecule lagged behind the observations, forcing researchers and doctors to rely on crude extracts of cells for study. For years, Fleming remained involved in antibiotic research and was eventually able to isolate the molecule that killed *S. aureus.* He named the chemical "penicillin," after the type of mold that grew on the Petri dish.[17]

Although this was a promising discovery, it took nearly a decade to find a reliable way to isolate enough penicillin to conduct experiments on its ability to fight bacterial infections in animals, and nearly another decade to mass-produce it. By the 1940s, the United States was the world's leading producer of penicillin, which found its way to the battlefields of Europe and Asia to treat wounded Allied soldiers suffering from infections. It's widely believed that penicillin was one of the most significant weapons the Allied forces had during World War II. The drug is estimated to have saved the lives of roughly a hundred thousand soldiers on the battlefront.[18]

The secret to penicillin's power to treat an infection comes from its ability to inhibit an essential bacterial enzyme. This enzyme, called transpeptidase, is a key builder of the bacterial cell wall, a strong mesh of sugar chains cross-linked by short proteins that provides structural support to the surface of the cell. Every time bacteria divide, the cell wall has to be expanded because the one bacterium grows before splitting into two. This is when transpeptidase is working the hardest. Penicillin, which finds its way into the depths of transpeptidase, jams the enzyme and stops it in its tracks, preventing bacterial division. Without the ability to divide, bacterial growth slows down, and the immune system of the patient is able to gain the upper hand in the fight. What makes penicillin so effective at treating infections is its specificity. Human cells don't have cell walls, so we don't have any enzymes that resemble transpeptidase. This enables penicillin to attack its target with pinpoint accuracy, without affecting human enzymes.

The development of penicillin ushered in a new era of "magic bullet" drugs—substances that can target a pathogen's enzyme without harming any of the enzymes belonging to the host. Scientists rejoiced at the thought of defeating every known infection and wiping out all disease-causing bacteria. But infectious diseases turned out to be more difficult to eradicate. Soon after penicillin became widely adopted, many bacteria began developing resistance to this seemingly magic drug, rendering it useless against more and more illnesses. The arms race was on.[19]

The bacteria's defense was again an enzyme. Through natural selection, new enzymes emerged that could chew up penicillin before it could come near transpeptidase. These enzymes, known simply as penicillinases, break down penicillin and many of its derivatives, giving bacteria a survival edge. With them, bacteria can withstand our chemical attacks, thrive, and proliferate. This has led to the emergence of worldwide antibiotic resistance. As more and more antibiotics are produced, new enzymes emerge to dismantle them, break them apart, or simply pump them out of bacterial cells. In the United States, these tiny enzymes cause the deaths of thirty-five thousand people a year from drug-resistant infections. Just as enzymes are crucial to our chemical warfare against bacteria, enzymes are microbes' main defense against our attacks.

A CATALYST FOR CHANGE

Gretchen Anderson was never as famous as Maud Menten. She would joke that one day she would win a Nobel Prize, but she knew her life-long calling was teaching biochemistry at a small school in South Bend, Indiana. Many students didn't have food. Some were unable to attend class because they couldn't afford gas. Gretchen committed herself to helping students flourish. She was the spark for everyone who met her, encouraging them to be curious and kind and push through adversity.

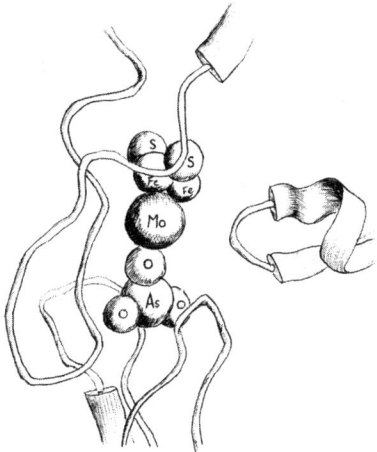

Arsenite oxidase (PDB code 8CFF) is an enzyme used by some soil bacteria to detoxify an arsenic-containing compound known as arsenite. A closeup of the active site shows how a molybdenum atom (Mo), along with iron (Fe) and sulfur (S) atoms, come into close proximity to the arsenite molecule (As and O) to facilitate the reaction. Here, molybdenum acts as a cofactor, providing the chemical properties needed for the enzyme to carry out the reaction.

But Gretchen was truly in her element when she talked about enzymes. In the research lab, her eyes sparkled and her smile was contagious as she spoke about arsenite oxidase, the enzyme a stinky soil bacterium uses to detoxify arsenic.[20] Above her desk on the wall hung a clock with the letters Mo, the symbol for an obscure element known as molybdenum. Sitting right at the center of the periodic table, molybdenum is a heavy metal often used for making alloys, flame retardants, and smoke suppressants. But molybdenum also happens to be essential for the function of arsenite oxidase. Deep inside the enzyme, amino acids surround a molybdenum atom, which the bacteria has scavenged from the soil and placed near a tiny cube made of iron and sulfur. The arrangement of the molybdenum, iron, and sulfur inside the enzyme is the key to its ability to carry out the reaction that turns arsenite into a less toxic form known as arsenate. The molybdenum and the clusters of iron and sulfur act as a sink for electrons from arsenite, converting it to arsenate. The molybdenum metal in this case acts as a "cofactor," a nonprotein component that helps an enzyme do its job.

It might seem strange for an enzyme to use such an obscure metal, but cofactors are not unique to arsenite oxidase. Many enzymes, including some of our own, use these nonprotein components to help them function properly.[21] The twenty different amino acids exhibit a range of chemical properties and can be sufficient for most proteins to function, but sometimes enzymes need more help. For many enzymes, elements like molybdenum, iron, and sulfur provide this added chemical diversity, expanding the types of reactions that enzymes can accomplish. In a sense, the amino acids of many enzymes provide a framework for holding these cofactors in place. The precise geometry created by all three components—substrate, enzyme, and cofactor—provide everything that is needed to turn a substrate into a product.

Many metals act as cofactors. Iron, zinc, magnesium, and cobalt are a few examples of essential trace metals that we need to have in our diet for our enzymes to function properly.[22] And like *Alcaligenes faecalis,* we use molybdenum in our enzymes to detoxify sulfites and to recycle excess DNA and RNA bases.[23] This cooperation between proteins and metals, whereby proteins capture the unique reactive properties of metals to carry out essential biological functions, is found across all known species. Some of these partnerships are truly remarkable. For instance, a species of extreme archaea known as *Pyrococcus furiosus*, literally translated as furious fireball, lives near hydrothermal vents at temperatures approaching the boiling point of water. Enzymes isolated from this extremophile have been found to use tungsten, the same element used to make the filaments inside light bulbs—as a cofactor.[24]

Metals are not the only types of cofactors used by enzymes. In fact, many of the vitamins we consume in our food or in a multivitamin gummy are cofactors for a lot of our enzymes. For example, vitamin C is needed by enzymes that build collagen fibers, ultimately forming the strong web surrounding our cells.[25] Collagen is part of the extracellular matrix that prevents our cells from bursting. Without access to vitamin C, these enzymes will not function properly, in leading to the bleeding gums and poor wound healing that can result as a condition known as scurvy. Vitamin K is needed for enzymes that activate the clotting response, and its deficiency can cause excessive bleeding.[26] Folic acid, another name for vitamin B9, is an essential cofactor for enzymes that make the A and G bases in DNA. Because of the role it plays in DNA synthesis, folic acid is important for cell division, especially during development. A lack of folic acid during pregnancy can cause abnormalities in fetal development because the growing fetus can't make enough DNA bases to keep up with its rapidly dividing cells.[27]

In her class, Gretchen explained that an enzyme provides the spark that helps a chemical reaction proceed. Even the most spontaneous reactions face an energy barrier that must be overcome. For example, burning wood in a fireplace is a spontaneous reaction, but it still requires a spark or a small flame to get it going. Enzymes provide the push needed to overcome those energy barriers, and once the spark is lit, the reaction can occur. An enzyme does this by twisting and turning the substrate, contorting it into uncomfortable positions that it would otherwise not easily adopt (and sometimes with a little help from a metal or vitamin cofactor). Now the molecule is no longer a substrate;

it is the product of a chemical reaction, and at this point the enzyme releases its grip and allows the newly made product to diffuse into the surroundings.

Gretchen died in December 2019. She had been diagnosed with breast cancer the year before. Life had never manifested in a person so fully as it did in Gretchen. It seemed everything she touched, every word she spoke, catalyzed energy in everything around her. At her funeral, not a single person mentioned her scientific achievements. There was no mention of arsenite oxidase or molybdenum, her list of publications, or her scientific accomplishments. Instead, we all bore witness to her love of life, which she made sure lived on in all of us. The pastor ended Gretchen's memorial by speaking of a fundamental law of physics—that energy cannot be destroyed; rather, it changes form. The energy that is Gretchen can never be gone because science will not allow it. She lives on through how she changed all of us.

People like Gretchen are like enzymes. They find you and see your potential for change. They bring you in and shield you from the surroundings. In doing so, they provide you with the right environment for change: the environment you need to realize your own potential. They help you overcome energy barriers; they are the spark you needed. Even if the process contorts you and makes you uncomfortable at times, the positions they put you in transform you. Then, when you are ready, they release you as a new form of yourself—changed, rearranged. We are neither created nor destroyed, but rather recreated. It all happens spontaneously. And like a true catalyst, they go on to change the lives of the next person and the next, doing it all over again, unphased, unchanged in the process.

Just as the catalysts in our lives have the capacity to change the world by changing people, one person at a time, the tiny machines we call enzymes, discovered less than 150 years ago, also trigger changes on so many scales. They control how plants respond to the changing seasons and when and how bacteria divide. They help viruses invade our cells while our own enzymes fight the viral infection on our behalf. Some are natural killers, like those found in the venom of snakes. Others give life, like the photosynthesis enzymes that use sunlight to capture carbon dioxide, turning it into food for nearly all organisms on earth.

NEW ENZYMES, NEW ERAS

It is not an exaggeration to say that enzymes have changed the face of our planet, and perhaps more than once. Starting at the dawn of life, ancient enzymes made life possible, and new ones transformed entire ecosystems, determining the rise and fall of scores of species and even influencing Earth's climate. In fact, many of the transitions between the different eras in the history of our planet took place because of changes on the molecular level, when old proteins learned new tricks. The new enzymes created diversity and expanded the reach of organisms to all corners of the world. Around the time that amphibians were venturing onto dry land, new plant enzymes were beginning to emerge and carve

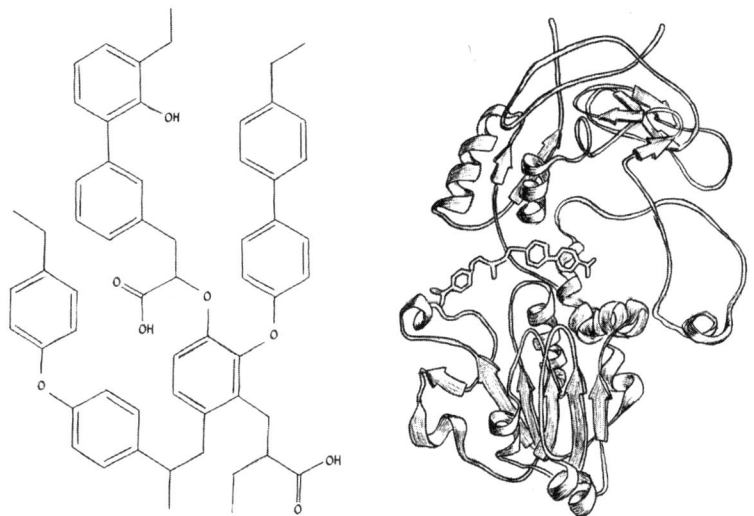

Lignin is one of the main components of bark. The multi-ring structure provides strength and rigidity, allowing trees to grow to great heights. Many enzymes are required for the assembly of the individual rings into the large networks of lignin. One of these enzymes is cinnamyl alcohol dehydrogenase, shown here bound to an intermediate compound containing three rings (PDB code 2CF6).

out a pathway for the creation of biomaterials the world had never seen before. About 370 million years ago, some plants found new uses for old enzymes by making modifications to the DNA that coded for them. These newly modified enzymes were destined to change the face of the planet. They made use of molecules that plants had used for millions of years as a type of sunscreen. These molecules, known as phenylpropanoids, are made of hexagonal rings of carbon, decorated in various ways by small modifications. Such arrangements act as powerful sun blockers by collecting harmful UV rays and dampening their damaging effects. As the new enzymes emerged, they began to connect the hexagonal rings together in different ways, constructing a strong web of carbons resembling a beehive. The arrangement, repeated over and over, began to cover the outside of plant cells, providing additional strength. These new enzymes had invented lignin, the main component of the wood found in the trunks and bark of modern trees.[28]

Armed with the solid scaffold of lignin, plants began to grow taller and stronger. No longer limited to waist height, the plants began to tower above the landscape, ushering in a new era on Earth. Small malleable plants became full-fledged trees with strong trunks and thick bark. As the trees grew taller, they gathered more carbon dioxide from the air to satisfy the new enzymes' voracious appetite for the carbon needed to make the lignin in wood. Billions of tons of carbon dioxide were siphoned from the atmosphere and became trapped in trees. This carbon capture marked the end of the Devonian era and began what's known as the Carboniferous era, a 60-million-year period during which carbon was sucked from the air at unprecedented levels and stored in giant tree trunks. As the trees absorbed more carbon dioxide, they released massive amounts of oxygen as a byproduct. This global event triggered powerful changes within the climate. With more oxygen, animals that were newly emerging from the water grew bigger and began to spread out on land. Dragonflies with two-foot wingspans hovered high above the trees, and new reptiles that lived exclusively on land became the first dinosaurs. All these global changes were possible because of the handful of enzymes responsible for making wood in trees.

Today, when a tree dies or is cut down, bacteria and fungi use their enzymes to digest the wood, extracting minerals and carbon from the extensive network of lignin within the dead tree. But when the wood-making enzymes first emerged on the planet, bacteria and fungi had not yet developed the tools to digest this tough new biomaterial that plants had stumbled upon. Nature, it seems, had figured out a way to make wood but had no means to break it down.[29] Dead trees simply did not rot. Instead, massive trunks and branches accumulated as each giant tree toppled over, piling up on top of each other. With the immense weight of the countless trees, the dead wood began to sink and became buried in the soil below. The intense pressure slowly changed the chemical composition of lignin over millions of years.

Today, these ancient burial sites of undigested trees can be seen in the form of coal and other types of fossil fuels: in fact, we extract millions of tons of prehistoric dead trees each year to satisfy society's need for energy. By burning these relics of the ancient world, we release back into the atmosphere the very same carbon that was absorbed hundreds of millions of years ago, and with it, we tip the scale toward a warmer climate. It took about 60 million years for the first enzymes that could digest the lignin in wood to emerge. Fungi and bacteria eventually found ways to break down the tough biomaterial that had given plants a huge survival edge. Today, the enzymes made by fungi and bacteria maintain a type of balance between the living and the dead. Living trees collect carbon dioxide from the air as they grow and pack it into the lignin in their trunks and branches. And when these trees die, the wood-digesting enzymes made by microorganisms put the nutrients back into circulation, returning minerals to the soil and carbon into the atmosphere.

The story of wood is the story of enzymes and a planet that both influences, and is influenced by, the life it shelters. And it is not only wood, and tiny enzyme machines, that star in the larger story of change on Earth. Today humans, like trees, soar to new heights. We create plastics and Styrofoam and other materials that nature cannot break down. Like the ancient trees, plastics will not rot, and Styrofoam will not decay, so these materials can't be recycled by our natural enzyme machines.

Every day, we create mountains of trash that litter the Earth. Like our tree ancestors, we grew too big for our own weight, and we crowded the planet with our newly made materials. Like the trees, we did not mean to leave our waste everywhere. But without a way to break down our waste materials, what will become of us? Will future civilizations millions of years in the future find massive piles of plastic relics?

The race to find ways to deal with the problem of accumulating plastic waste began decades ago, yet little has changed. If anything, we are producing more waste than ever before. As the population explodes, landfills grow bigger with unimaginably large mounds of plastics. Yet enzymes may once again offer hope. The answer to our waste problem might come in the form of enzymes that can potentially begin to digest the billions of tons of plastics littering our planet. And luckily, we don't have to wait 60 million years for these enzymes to emerge; scientists have already identified several enzymes that can chew up different types of plastics.

While inspecting a beehive, Federica Bertocchini, a professor at the University of Cantabria in Spain, spotted a number of wax worms. Bertocchini, an amateur beekeeper, recognized the worms as a common pest that can plague bee populations, so she immediately removed the worms and placed them in a plastic bag. As she was about to discard the bag, she noticed tiny holes and suspected that the worms had chewed their way out of it. Curious about what could break through the plastic, Bertocchini and her colleagues isolated two enzymes from the saliva of the worm and showed that they could break down polyethylene. This was a major discovery, since polyethylene accounts for roughly 40 percent of all plastics and is a major contributor to plastic waste.[30] With this discovery, scientists have been working not only to make these two enzymes more efficient at extracting carbon from plastic waste but also to engineer new enzymes that can degrade other types of plastics. One day, we may be able to sort our plastics into different bins—not recycling bins but bioreactors in which enzymes designed to chew up each kind of plastic can work to break down the material. Instead of sitting in landfills, plastics may one day rot, decompose, become bio-degradable, and be returned to the Earth by new enzymes.

One day, like Gretchen and Maud Menten, we will be gone. Enzymes made by bacteria, fungi, and worms will feast on our bodies. These enzymes are nature's way of returning precious nutrients back into circulation, where they can be taken up by other organisms. It seems that enzymes, like those caring, innovative mentors who are no longer with us, show us how life and death are entwined—how one life has the power to affect change in others.

6

TRANSFORMATION

MAGGIE M. FINK

MY HOME IN INDIANA SITS at the edge of a small woods. Far away from any city lights, my backyard becomes delightfully dark as the sun sets, especially at the peak of summer, when the trees are at their fullest. No light can penetrate the wall of limbs and leaves that cloak the garden, and my boys' well-used trampoline, in darkness. In the approaching night, where the only sound is the rustle of the leaves and the distant call of coyotes, the landscape comes alive. The woods twinkle with a gentle, rhythmic light, a flicker and glow that seem to telegraph secrets from a time when all beings spoke the same language. This dancing lightshow marks the arrival of the lightning bugs. You would call them fireflies if you weren't from the Midwest. Against the backdrop of darkness, these floating bugs flash in and out of sync, mimicking the soft glow of candlelight. I send my children outside to catch them, showing them how to cradle the fragile creatures gently in their palms, letting the lightning bug walk along the crevices of their fingers. I show them how to bring their cupped hands to their eyes, to peer through their fingers to see the flashing belly.

An animal that produces its own light is transcendental. It tugs at a longing we have as children for illuminating the darkness and summons a visceral feeling of awe, one that dates back to the origin of our own consciousness and the stories of our creation. In the opening chapter of the Bible, the world is described as void, without form. Darkness hovered over the face of the deep. And then there was light. Even before the sun, there was light. Light transforms. Light is powerful. Many ancient deities were associated with the sun. Perhaps it is our human-centric way of understanding the world, as creatures who rely on light above all other senses, to attach such importance to light. But throughout the natural world, we see creatures making their own light.

Not by discovering electricity, as humans did, but by using chemistry and proteins in their bodies to glow in the darkness. This phenomenon, called bioluminescence, reveals itself in the depths of the ocean, the hush of the night, and the hum of the rainforest.

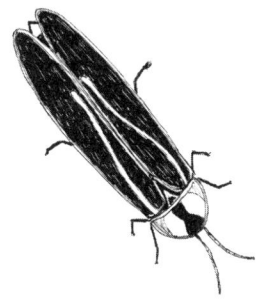

Fireflies produce a glow that is produced in the lower abdomen via a chemical reaction involving luciferin, oxygen, and the enzyme luciferase.

The coordinated lightshows of the firefly aren't just for our aesthetic pleasure; they are a language that transcends words, one that fireflies have developed to communicate with one another. For these nocturnal insects, light is the way they exchange signals of courtship and attraction: each flicker of light is a flirty, poetic declaration of interest. Fireflies also engage in synchronized displays, turning their collective glow into a mesmerizing spectacle that may help them identify members of their own species and may contribute to the intricacies of their reproductive rituals. The pulsating rhythm of firefly gatherings is a living testament to the complexity of communication in the natural world, where creatures can connect with flickers and flashes.

The glow of the firefly is also about survival. Fireflies contain chemicals known as lucibufagins, which are unpalatable and even toxic to certain predators, deterring them from eating the firefly. Many fireflies synthesize lucibufagins from chemicals they acquire in their diet; however, the female *Photuris* fireflies have evolved a cunning strategy to acquire these compounds, primarily by preying on other firefly species that already have lucibufagins.[1] Upon consuming their prey, the femme fatale fireflies extract and store the lucibufagins within their bodies, effectively incorporating them into their own defensive arsenal. This adaptation not only offers *Photuris* fireflies protection against predators, but also potentially enhances their survival prospects. Regardless of how fireflies make or obtain lucibufagins, the bright glow communicates the firefly's chemical defenses to those who might consider it a potential meal.[2] Firefly larvae also harness the power of bioluminescence. In their larval stage, fireflies emit a glow to attract prey. These voracious predators lurk in the leaf litter and soil, and their bioluminescent glow serves as an alluring trap for unsuspecting beetles, snails, and other small invertebrates. It's a strategy that blends the allure of light with the art of predation.

Yet the story of firefly bioluminescence doesn't end there. The glow also plays a crucial role in the firefly's navigation of its surroundings. As these insects flit through the night, the rhythmic patterns of their glow help them orient themselves and avoid collisions. Their light is a celestial GPS system crafted by evolution, a built-in feature that enables fireflies to navigate with precision.[3] The enchanting lives of fireflies tell a remarkable story of evolutionary marvels and ecological finesse, one in which glowing and flashing can be a means of communication, defense, predation, and navigation.

Far from the warmth of the Indiana July sky, with its cottonwood fluff and hovering living light bulbs, another form of bioluminescence thrives. In the Pacific Ocean, a tiny bobtail squid spreads its tentacles, emitting a mysterious light as it searches for prey. Each night, the bobtail squid uses bioluminescence from a light organ in its belly to disguise its silhouette from potential predators, and each morning, with the return of the sun, the light fades. This cycle is controlled by bacteria that live inside the squid—*Vibrio fischeri*.[4] Just after a bobtail squid hatches, *V. fischeri* from the ocean waters find their way to the light organ within the baby squid. These bacteria grow during the day and then, at night, once their numbers have increased enough, they glow all at once, enabling the bobtail to camouflage with its environment as it searches for food. The light organ has specialized filters that diffuse the light to match the moon's glow and erase the bobtail's shadow on the ocean floor. This counterillumination protects the squid and, in turn, provides the bacteria with a safe place to colonize. As night passes, the squid and its bacteria continue to work together. *V. fischeri* continues to produce light.[5] But when the sun begins to come up, the bobtail squid ejects nearly all the bacteria back into the ocean, saving just a tiny fraction inside its light organ. This essentially turns off the lights. Then, as the squid buries itself in the sand for the day, the few remaining bacteria begin to divide and grow safely inside it.[6] And the process is repeated. The sun begins to set, the squid emerges from the sand, and *V. fischeri* again turn on the lights.

The glowing lights of *V. fischeri* are not a new discovery. Bioluminescence shows up in some of the earliest human records. Mariners of ancient times spoke of a "fire in the sea." The ancient Greeks, in particular, marveled at the sparkling waves as they sailed across the Mediterranean. They attributed this luminous spectacle to the divine, believing it to be a gift from the gods. In the fourth century BCE, the Greek philosopher and naturalist Aristotle was among the first to document the glow of marine creatures in his treatise *Historia Animalium*. He marveled at the enchanting blue-green lights emitted by jellyfish and suggested that the radiant displays were a form of communication or defense. As science progressed and began to explain the natural world, the mystery of bioluminescence piqued the interest of many. In the seventeenth century, naturalists like Sir Francis Bacon and Robert Boyle began to study the ethereal glow emitted by some marine organisms, igniting a scientific curiosity that would shape our understanding of this phenomenon. Even Dutch scientist Antonie van Leeuwenhoek, famed for his work with microscopes, turned his lens toward bioluminescent organisms. He described the minuscule, radiant "animalcules" he observed in water as "living fireworks."[7]

But it wasn't until the late nineteenth century, in the dimly lit laboratory of Raphaël Dubois, a French pharmacist, that a mechanism and name was given to the phenomenon. The ability to investigate proteins and their functions did not yet exist. Scientists hardly knew what a protein was, let alone how it could generate light inside the belly of a beetle. Dubois set out to unmask the secrets behind this captivating phenomenon. At first, he meticulously dissected click beetles and fireflies and examined their tiny, glowing lanterns in his lab. But when neither organism yielded new insights, Dubois turned to bioluminescent clams.[8] He began by grinding the clams into a paste, which still produced a bright blue light. He then separated the paste in half. He placed one half into boiling water, which quickly eliminated the light, and left the other half alone. When he recombined the two halves, however, the light returned. He concluded that the light was not caused by a single component, but rather by the interaction of two separate molecules—an enzyme and another small molecule called a substrate. While many small molecules aren't affected by heat, nearly all proteins unfold and stop functioning when exposed to high temperatures. This is why the heated clam paste immediately ceased to glow. But because the unheated half still contained a functional enzyme that could carry out its chemical reaction on the small molecule, when the two halves were reunited, the glow returned. Eventually, as the enzyme used up all the small molecules, no more chemistry could be done, and the light slowly faded.

With his discovery, Dubois unveiled the mechanism of bioluminescence. He named the enzyme luciferase and the small molecule substrate luciferin, after Lucifer, the light-bearer in classical mythology. This dynamic duo is responsible for the ethereal radiance of not only clams but also click beetles, fireflies, and thousands of other organisms that can glow in the dark.

Chemical reactions require energy. Burning wood uses the chemical energy stored in wood fibers, which is released as thermal energy in the form of heat and light. When we clap our hands, we put in energy in the form of motion and force, which is then transformed into a sound. At the molecular level, energy is stored in chemical bonds. Breaking them releases energy, which can be harvested by cells to power all kinds of processes. This energy is precious, especially for tiny organisms. So for any animal, fungus, plant, or bacterium to take energy that could be used to keep itself alive and give it up to generate light instead indicates that light emission must be extremely important for survival in one way or another. There must be an intrinsic biological drive to make light, just as there is a drive to seek light.

The biological impetus for light production is rooted in its evolutionary advantages, which explain why bioluminescence is found in organisms of all shapes and sizes.[9] In marine environments, many species of fish use light for social interactions and coordination within their communities. This communication strategy aids in attracting mates and coordinating group activities in the vast, dark expanses of the ocean water or in crowded shallow waters. Bioluminescence is also a potent defense mechanism. Some organisms use light to deter predators or confuse them. The luciferase enzyme, for instance, is rapidly activated when an organism senses a threat. The resulting burst of light startles or distracts predators, offering the bioluminescent organism the chance to evade capture.

Despite how common glowing organisms are, the evolution of bioluminescence and luciferases remains unclear. With so many different organisms, from bacteria to fungi, that produce light, we might assume that luciferases come from a single common ancestor; with more and more research into the origin of luciferases, however, it has become apparent that even bioluminescent organisms with shared ancestry have evolved to produce light at different times and in different ways.[10]

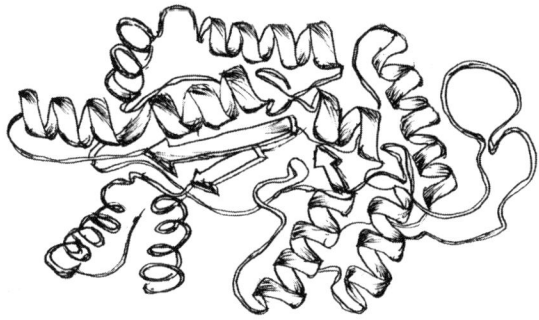

The luciferase protein from the firefly *Photinus pyralis* is responsible for the glowing belly seen on the firefly's underside (PDB code 1LCI).

Jing-Ke Weng, a chemist at Massachusetts Institute of Technology, recently set out to study the evolution of luciferases in beetles, which include the ubiquitous firefly. Specifically, he was interested in *Photinus pyralis*, commonly called the big dipper firefly, which is found across North America. First, he and the researchers in his lab tried to determine where the luciferase gene was located in the big dipper's DNA. Genes are often clustered together based on what biological task the resulting proteins will carry out. Weng's lab found that the big dipper luciferase gene was located in the middle of a cluster of four genes all coding for an enzyme called fatty-acid CoA ligase.[11] This enzyme helps assemble long chains of fatty acids to carry out diverse functions in the cell—functions unrelated to bioluminescence. Yet the luciferase gene and the fatty acid CoA ligase share a lot of similarities in both DNA and amino acid sequence, indicating that at one point in time, luciferases may have been a fatty-acid CoA ligase. Often, when identical genes show up all in a row like this, it is the result of a duplication event, of one copy of a gene turning into two copies during cell division. As duplication events occur, the additional copy of a gene can turn into a protein that, with just a few small changes, can have an entirely new purpose. In the case of the big dipper, the researchers predicted this is what happened with luciferase. A protein needed to make fatty acids had turned the big dipper into a glowing insect.

But understanding how luciferases arose in one organism doesn't fully answer the question of when this event of bioluminescence occurred in evolutionary history. To investigate this question, Weng and his team gathered DNA from two more glowing beetles: *Ignelater luminosus,* a species of click beetle native to Puerto Rico, and *Aquatica lateralis,* more commonly known as the Japanese Heike firefly. While the big dipper firefly and the click beetle are related, albeit distantly, their light organs are on opposite sides of their bodies. At the top of the click beetle's head sit two spots that light up just like the belly of the firefly. The Heike is more closely related to the big dipper than the click beetle; their evolutionary distance, however, is still around 100 million years, further apart than humans and rodents.[12] By investigating luciferase genes from these two additional beetles, Weng and his team hoped to determine whether these bioluminescent insects came from the same ancestor or if their light-making abilities developed independently, perhaps millions of years apart. They discovered that in the more closely related Heike firefly, the luciferase gene was found in an incredibly similar location to that of the big dipper, flanked by four copies of the fatty-acid CoA ligase. This indicated that these two fireflies gained their bioluminescence from the same ancestor. In the click beetle, however, luciferase had no ligases nearby, suggesting that bioluminescence had emerged via another evolutionary tree.

Aside from beetles, it is estimated that more than seven hundred genera of animals are bioluminescent.[13] With so many different luciferases, fully understanding how each one evolved is a nearly impossible feat; scientists believe, however, that there are at least ninety examples in nature of luciferases evolving independently, meaning there is no single common ancestor that the very first bioluminescence can be traced back to. This phenomenon of independent evolution is not unique to luciferases. In fact, the antifreeze proteins discussed in Chapter 1, which protect many organisms from the cold, emerged in a similar manner. Multiple proteins with very different shapes evolved separately to accomplish the same task. This type of convergent evolution is seen in many aspects of life, but none as mesmerizing as a glowing insect in the air or a squid hovering over the ocean floor. With around 80 percent of known bioluminescent organisms residing in the sea, a world of constant darkness, it is clear that bioluminescence favors specific environmental niches where light is particularly beneficial.[14] And given how much it pops up in nature, it seems to be an evolutionarily easy process, arising from changes in proteins where light is a lucky byproduct of other adaptations.

Convergent evolution has produced many luciferases that don't share a common structure, despite accomplishing the same biochemical reaction. This is in contrast with most other proteins, where a common structure is used to carry out similar functions in different organisms. For example, structural proteins like human microtubules or spider silk, which have the same basic job—creating strength and structure—have similar shapes across all forms of life. As organisms evolve, proteins are repurposed, using the same structure to accomplish a new function. But in the case of luciferase, it is not the architecture of the protein but rather the chemistry that is preserved. For most enzymes, just a few amino acids are responsible for carrying out a chemical reaction. The rest of the enzyme is composed of amino acids that help maintain its three-dimensional structure and provide regions for other molecules such as oxygen, sugars, free amino acids, metals, and even other proteins to interact with it. As these molecules come into contact with luciferase, they can introduce structural changes that alter how the enzyme is able to latch onto the actual substrate. Sometimes when these molecules bind, they can inhibit luciferase, either stopping its activity or slowing it down. Other times, they can cause the enzyme to work faster, depending on where they stick to the structure. Because the environment can change rapidly, having sections of an enzyme devoted to controlling its activity can help an organism fine-tune its molecular response. By keeping the amino acids that actually perform the chemistry and allowing changes everywhere else, an enzyme can more readily evolve in response to specific environmental cues.

Presently, twelve different types of luciferases have been identified, most found in specific groups of bacteria and animals.[15] Despite a limited understanding of how bioluminescence evolved, as far as scientists know, all luciferases share a common feature—they consume oxygen. The universal need for oxygen hints at an underlying function that may have had nothing to do with creating light. Indeed, many researchers have proposed that this extraordinary ability to produce light may have evolved to eliminate oxygen or reactive oxygen species (ROS) from organisms.[16] ROS are tiny, highly charged particles that behave like chemical sparks and can damage DNA and proteins if left unchecked. Oxygen, while essential for life, can be a double-edged sword, giving rise to potentially harmful ROS during biological processes such as metabolism and even protein folding. Bioluminescence, by harnessing the energy released as luciferin is transformed, may have evolved as a biochemical strategy to detoxify organisms by using oxygen in a controlled manner. Many organisms contain proteins similar to luciferases that act on luciferin-like molecules but produce no light, also pointing to the hypothesis that bioluminescence was a lucky accident. The notion that this captivating phenomenon might have evolved as a mechanism to eliminate oxygen or ROS paints a vivid picture of the intricate dance of life. Luciferases not only illuminate the dark depths of our oceans, but also shed light on the ingenious ways in which organisms have evolved to navigate the challenges presented by their environments. Bioluminescence is a testament to the remarkable adaptability and ingenuity of life on our planet.

LIGHT-UP LUCIFERINS

In the muggy forests of South America, we find one of nature's most impressive examples of bioluminescence and the role that luciferins play in creating light shows. The railroad worm, a beetle larva known as *Phrixothrix hirtus*, has long intrigued scientists with its unique bioluminescent display. The worm gets its name from the pairs of tiny glowing lights that run along its body, which evoke the image of a train traveling through the dark. What sets the railroad worm apart from other bioluminescent organisms, however, is its ability to produce two different colors of light: a reddish-orange light from its head and a bright yellow-green light along its body. The exact mechanism behind this color variation is not fully understood, but it's believed to involve subtle differences between luciferases found within the beetle larva's head and those in its body.[17] The main difference between these luciferases is the size of the binding pocket, a hollowed-out spot on the enzyme where a luciferin molecule can easily fit. For the railroad worm, the luciferase along the body has a smaller binding pocket, forcing the luciferin to fit in tightly. The cramped environment presses together the oxygens and hydrogens in the binding pocket and on the luciferin, causing the electrons to repel each other. As luciferase transforms luciferin, these tight interactions are finally released as energy in the form of a green-yellow light. The energy needed to emit this specific color is much higher than what is required for the red glow at the top of the beetle's head. The binding pocket of the luciferases localized there is more open. This allows the luciferin to lazily settle in with much less of the repelling energy that occurs along the back. When the luciferase and luciferin perform their chemical dance in this binding pocket, the energy released is lower than that of the green-yellow output, resulting in red light and giving this railroad worm its signature dual-color bioluminescence.

If luciferases are the protein engines of bioluminescence and the catalysts that drive the conversion of chemistry into light, then luciferins are the fuel needed for the reaction. All luciferases have a luciferin

Luciferin structures can vary from species to species. Many luciferins from different species incorporate similar structures such as heterocyclic rings, including some species of crustaceans *(top)* fireflies *(second from top)*, and sea pansies *(third from top)*. However, bacterial luciferins *(bottom)* are simply a chain of carbons and hydrogens.

partner that they act on, breaking and creating chemical bonds to harness and release energy as light. Like luciferases, the structure of the luciferin molecule can vary wildly from organism to organism. Some are simple and elegant, existing as a line of carbons, while others are constructed from more intricate rings decorated with nitrogens, sulfurs, and oxygens.

The largest class of luciferins is the coelenterazine family, widely distributed among marine organisms such as jellyfish, corals, and certain species of plankton. Coelenterazine-based bioluminescence involves a chemical reaction that incorporates molecular oxygen into the luciferin structure to produce light. In the center of the coelenterazine luciferin structure lies a group of atoms known as the imidazopyrazinone, consisting of conjoined five-membered and six-membered rings containing three nitrogen atoms.[18] This ring provides stability to the luciferin molecule while also facilitating the chemical transformations required for bioluminescence.

A distinguishing feature of coelenterazine luciferins is their versatility in producing a spectrum of colors in bioluminescent displays. The variation in color is attributed to modifications in the chemical side chains attached to the imidazopyrazinone core. These side chains allow the addition of groups of chemicals, which can include amines, hydroxyls, or other combinations of hydrogens, oxygens, and carbons.[19] This adaptability allows organisms that use coelenterazine luciferins to tailor their luminous displays to specific ecological needs, such as attracting mates or deterring predators.

Despite this diversity of structures, most marine creatures that rely on coelenterazine luciferins can't synthesize them on their own; instead they acquire them through their diet.[20] In the ocean's food chain, organisms at different levels contribute to the dissemination of coelenterazine luciferins as they pass through the digestive systems of predators and prey alike. The journey of these bioluminescent compounds begins with the consumption of organisms that either produce or accumulate coelenterazine luciferins as part of their biochemical repertoire. Many jellyfish are renowned producers of coelenterazine luciferins, and their incorporation into the diet of various marine organisms establishes a transfer of bioluminescence across the food chain. As small fish, crustaceans, or other marine organisms consume jellyfish or plankton that contain coelenterazine luciferins, these compounds become assimilated into the tissues of the consuming organisms and can make them glow, too.

Other types of luciferins, including the beetle luciferin, have evolved structures that contain a set of carbon rings and sulfurs.[21] And some, like the bacterial luciferin, have no rings, just a stretch of carbons and hydrogens punctuated by oxygen.[22] Like their luciferase counterparts, the evolution of these structures in bioluminescent organisms remains elusive. Even the question of how different organisms make their luciferins has been fully worked out only for bacterial and fungal species. In each of these species, the genes encoding luciferins are grouped together in a section of DNA. In bacteria, this set of genes is called the *lux* operon. Only small differences in the resulting luciferins have been identified between different bacterial species, most of them simple adjustments to help the bacteria adapt to their environments. Regardless of how bacteria, fungi, jellyfish, or fireflies make their own luciferins, evolution has sculpted these biochemical pathways to allow each organism to harness light in the most resourceful and adaptable ways. Whether flickering in the depths of the ocean or illuminating the summer night, the biosynthesis of luciferins unveils the hidden wonders of nature's biochemical repertoire.

THE BENEFITS OF SYNCHRONY

We have some idea, then, about how fireflies create their glow, at least from the perspective of a single luciferase enzyme. But how does every luciferase in the firefly know when to grab onto a luciferin and turn the lights on simultaneously? Inside a cell, proteins and molecules of all shapes and sizes bounce around, each one trying to accomplish its task. Without coordination, luciferases would be out of sync and produce

only sporadic, dim light. To ensure that light generated from the luciferase reaction is synchronized, producing a flash of light at a precise
moment, billions of luciferase enzymes must act at the same time. Many
organisms, then, have developed sophisticated regulation mechanisms
to ensure that when it's time to glow, luciferases are activated in such a
great number that every available luciferin and luciferase can meet up
to create the brightest light possible.

The regulatory systems that govern the actions of proteins in a cell
can be complex and vary dramatically, not only between organisms but
even between different types of proteins. One reason for employing
these regulatory mechanisms is that it can be energetically expensive
for a cell to make proteins. Energy is required to assemble amino acids
one at a time, and often, many different proteins are required to make
and process the molecules needed for the organism's survival. Without
the ability to efficiently regulate when these proteins are made, it would
not be able to function. For a firefly, too, regulation is critical not only
from an energy standpoint: the synchronization of luciferase activity is
also essential for finding a mate at the right time and ultimately, for the
survival of the species. Similarly, other organisms like the bobtail squid
and its companion, V. fischeri, need regulation to synchronize when luciferases are produced and active.[23]

Regulation mechanisms can take many forms. Some of the most
well-studied mechanisms happen at the DNA level. Remember the central dogma of molecular biology, that DNA, the instruction manual, is
transcribed into RNA, which is then translated into a protein. But what
if the DNA becomes inaccessible? Cells have evolved ways to physically
block DNA to keep a protein from being pieced together. Proteins called
transcriptional regulators act as molecular switches for converting DNA
information into functional proteins. These regulator proteins latch
onto specific DNA sequences they recognize. Some of these regulators
act as a barrier, effectively keeping that genetic material "turned off" and
repressing transcription. Others actively recruit the transcription machinery, encouraging more production of whatever the gene contains.
Often, these repressors or activators are specific to certain genes or

groups of genes, but there are some that regulate the transcription of a wide range of genes. The environment inside and outside of a cell can provide information to these regulators, allowing for the appropriate proteins to be made in response to environmental cues.

One of the most basic examples of regulatory proteins that every biochemist is taught is called the *lac* repressor. First discovered in *E. coli*, the *lac* repressor protein binds to DNA just before a series of genes that encode for proteins that digest the sugar lactose, which *E. coli* metabolizes for energy. When there is no lactose in the environment, there is no need to make the proteins that digest lactose—it would be energetically costly to make proteins that can't be used, so the *lac* repressor ensures that none are made.[24] But if at some point *E. coli* encounters lactose, it will need to make those proteins. So if lactose begins to enter the cell, the *lac* repressor senses it and lets go of the DNA, allowing transcription to occur. This simple mechanism is especially common in bacteria, but multicellular organisms employ similar types of regulation, in addition to more complicated protein complexes that are still being uncovered and characterized.

Even light itself can be a cue to turn genes on and off, specifically the potentially damaging ultraviolet (UV) radiation that emanates from the sun. As sunlight touches our skin and penetrates our cells, it can cause DNA damage. But cells have a remarkable system in place to sense UV radiation and spring into action to repair the damage. This starts with a set of proteins that, like tiny sentinels, detect the UV radiation and send out an alarm signal. The signal triggers a cascade of events that activate specific genes and instruct the cellular machinery to produce the necessary proteins to fix the damaged DNA. It's like a well-coordinated dance, with each step carefully choreographed to ensure the health and survival of the cell.

Sometimes it isn't practical for a cell to regulate when proteins are made at the DNA level. For the firefly, the luciferases are all made and ready to go, sequestered in specialized cells called photocytes.[25] But the firefly still needs to regulate when they glow and flash. It has no time to wait for the luciferases to be made—potential mates won't be so patient. Because luciferases can't produce light without oxygen, the quickest way to turn on the lights is to control the flow of oxygen into the light organ. When the time is right, the firefly opens up a channel to flood the waiting luciferases with oxygen from the air, and immediately, the belly starts to glow. As one firefly initiates this light pulse, others nearby catch the beat and join in, creating a harmonious symphony of flickering lights. The mechanism behind this synchronization is still a subject of ongoing scientific investigation, but it likely involves a combination of environmental cues, genetic predispositions, and perhaps even an insectile form of communication. This simple mechanism of regulating luciferase activity happens very quickly from a human perspective, but that is precious time that a firefly needs to attract a mate or escape a predator.

Studying the regulation of these luciferase proteins provides insights into how our own proteins are regulated. When we eat, digestive enzymes are turned on, often by snipping off a segment of amino acids that renders them inactive. Proteins, like clotting factors, have a similar mechanism. If you cut your finger while slicing onions in the kitchen, your body has little time to stop the bleeding. Instead of waiting for clotting factors to be made from scratch, the proteins are already made and waiting on reserve, but inactive. When a small segment of them is sliced off by activating enzymes, clotting factors immediately go to work recruiting numerous molecules to constrict blood flow and eventually build a scab over the wound. Clotting factors, of which there arc twelve, are primarily synthesized in the liver and circulate in an inactive form within the bloodstream.[26] Each clotting factor plays a unique role in the clotting process. Activating the clotting response requires large complexes made up of multiple enzymes that convert the inactive clotting factors into their final active forms. This ensures a rapid response to injury while maintaining precise control over the coagulation cascade. The intricate interplay of activation and inhibition orchestrated by these proteins safeguards the delicate equilibrium between preventing unwanted clots and stopping bleeding as quickly as possible.

From bioluminescence to clotting factors, the ability of every cell in every animal to regulate its resources and energy in response to the environment is a remarkable, and still unfolding, story. With every new mechanism of control that scientists uncover, regulation emerges as essential to this narrative of balance and adaptation, from the DNA level all the way up to where proteins are stored in a cell. The relationship between the environment and the millions of decisions a cell must make every moment is not yet fully understood, but science is evolving, too. While big questions of protein regulation and even luciferase and luciferin evolution seem out of reach for now, it might take only another dive to the bottom of the oceans or a trek deep into a forest to find the answers.

NOT ALL THAT GLITTERS

Somewhere deep in the Atlantic, a ghostlike presence moves slowly and mysteriously through the dark waters. Its shape resembles a spaceship from a 1950s sci-fi novel—rounded at the edges with scattered lights, flashing as it moves, as if beaming a secretly coded message to another galaxy. In the darkness of the ocean, *Aequorea victoria* flaps its tentacles ever so lightly as it swims. This type of familiar jellyfish belongs to one of the most ancient groups of species alive today. Nearly 500 million years ago, ancient comb jellies split from our own ancestors. Today, species of jellyfish, including *Aequorea victoria*, roam the ocean, producing a tantalizing green light.[27] But unlike fireflies, railroad worms, or the bacteria inhabiting the bobtail squid, the jellyfish has a unique way of shining. Instead of using bioluminescence, the jellyfish emits green light using a protein known as green fluorescent protein or GFP. Unlike the luciferases found in bioluminescent organisms, GFP is not an enzyme; it does not break down chemical bonds to produce light energy. Rather, it uses light itself. It absorbs light at a certain color and emits a different color. Inside the jellyfish, two proteins sit side by side. The first protein, *aequorin*, scavenges calcium from the ocean and produces invisible rays of ultraviolet light.[28] The GFP nearby collects the

UV light and uses its energy to produce a green glow that can be seen on a starry night swim.

The key to the fluorescent property of GFP lies in its unique barrel shape.[29] Eleven beta strands twist around the outside of the barrel with amino acids pointing in and out of the barrel. Deep inside the barrel, three amino acids—tyrosine, glycine, and serine—come together to form the fluorescent core. Alone, none of these amino acids could produce fluorescence in the visible region of the light spectrum. But together, they form a new connection that transforms them into a fluorescent molecule capable of absorbing UV light and emitting it back as green light. This unique mechanism requires no enzymatic activity and no additional external factors once the three amino acids join. The genetic information encoded in the gene for GFP contains all the information needed for the protein to form the barrel structure and its fluorescent core.

Green fluorescent protein (GFP) was originally derived from the jellyfish *Aequorea victoria* and is responsible for its green glow.

The ability of GFP to glow green under blue or UV light was first discovered by the Japanese scientist Osamu Shimomura in the 1960s.[30] As a teenager, Shimomura survived the barbaric atomic bomb attack by the United States on Nagasaki, not far from where he worked in a munition factory. Always curious about the world, he began studying chemistry, landing a position in the laboratory of Yoshimasa Hirata at Nagoya University, despite not having a high-school diploma. His first task was to isolate luciferin from *Cypridina*, a tiny glow-in-the-dark marine crustacean. This was Shimomura's introduction to the world of living light. With his successful isolation of luciferin from *Cypridina*, he received an invitation to join a team at Princeton, where scientists had begun working on the molecular basis of jellyfish glow.

For months, Shimomura, his wife, and several research students traveled to the West Coast of the United States to collect jellyfish for their experiments. Then Shimomura spent countless hours dissecting the jellyfish's glow ring, a ring of organs that emit a blue color around the umbrella of the body, while trying to isolate the luciferin and luciferase that were responsible for the jellyfish's bioluminescence. After many failed experiments, Shimomura began to wonder if they had been wrong to assume that a luciferase enzyme had to be responsible for this glowing creature. Then one day, while testing the effect of pH on crude extracts of cells from the glow rings, he tossed a beaker of it into a sink that had sea water flowing into it. Immediately there was a flash of bright blue light as the extract touched the water. Convinced the seawater held the key to illumination, he tested each of the components of seawater and found that calcium was the molecule needed for the illusive protein to glow. This novel discovery pointed to a new type of light-emitting protein. One that needed only calcium.

It took over ten thousand jellyfish to get enough of this protein to take back to Princeton and study its mechanism further. The researchers named it aequorin and determined it was an enzyme that behaved very differently from the well-studied luciferase. While aequorin did bind to a luciferin molecule, it didn't emit any light until calcium also interacted with the protein, causing the structure to shift just enough for aequorin to transform the luciferin. This modification of the luciferin molecule released energy in the form of light, like other luciferases. Another key difference was how quickly this light could be emitted. As opposed to the slow glow that appeared from the click beetle and clams studied by Dubois, aequorin rapidly put out a flash of light when it sensed calcium, but took much longer to regenerate its ability to continue producing light. Aequorin was only the beginning of the story for Shimomura. The painstaking efforts to extract this new protein from jellyfish captured another, tagalong protein. Research into that other protein would ultimately lead Shimomura to win a Nobel Prize.[31]

As Shimomura and his team were extracting proteins from tens of thousands of dissected jellyfish, they noticed the presence of a green light in addition to the blue glow they had been studying. Eventually they determined this was also caused by a protein, though the protein was only present in extremely small amounts. Calling it green protein at first, they would spend nineteen summers catching jellyfish, transporting them across the United States, and cutting them up with scissors to get enough of this protein to even begin to study it. Unlike aequorin, green protein needed no other molecule to glow. Its light was not due to a chemical reaction. Instead, the green glow had to do with the blue light from aequorin. The energy of the blue light from aequorin was captured by green protein deep within its barrel-shaped core and transformed into a green glow. It seemed clear to Shimomura that green protein, eventually renamed GFP, was unique. Its glow was not produced by an enzyme like the bioluminescence of luciferases, but from something self-contained within its structure.

Scientists remained skeptical of Shimomura's claim well into the 1980s. Then, another researcher at the Woods Hole Laboratory in Massachusetts, Douglas Prasher, was able to isolate the gene that codes for the GFP protein in jellyfish. Today, thanks to advances in molecular biology, isolating a gene is fairly straightforward, but in the 1980s it required a herculean effort. It took months to isolate DNA from jellyfish, to cut it up into smaller pieces, and to search through the pieces to find the segment that contained the GFP gene. Prasher wondered if other organisms would be able to make the green glowing proteins if given the gene for GFP. If they could, that would prove that GFP is all that is required for fluorescence, as Shimomura had hypothesized. Prasher worked with two other scientists, Martin Chalfie and Roger Tsien, to introduce the gene to *E. coli* bacteria, using the bacteria's own protein-making machinery to make GFP.[32] At the time, introducing genes to *E. coli* was a relatively new technique, much more difficult and inefficient than it is today. Despite this, the initial work was successful, and the two scientists showed that bacteria will give off a green glow when the jellyfish GFP gene is provided to them. Later, they also showed that the same thing happens when a tiny worm known as *C. elegans* is given the gene for GFP, and that even plants could glow green from the protein. Prasher hoped that GFP could be used as a tracer that when fused with other proteins in a cell could help scientists track the location and movement of proteins under a microscope. But before he could see his dream come true, Prasher's funding ran out, and he had to leave academic research altogether.

In 2008, the Nobel Prize committee decided to award Shimomura, Chalfie, and Tsien the chemistry prize for their work in developing GFP as an essential tool for studying biological systems. There was no mention by the committee of Prasher, who at the time was working as a shuttle bus driver at a car dealership in Huntsville, Alabama. A stroke of bad luck and a failure to renew his funding had led Prasher to bounce from one job to another over the years while the gene he had cloned was being used by countless scientists as a valuable research tool. For example, by tagging GFP to the long fibers of microtubules in a cell, scientists could see the process of cell division happening in real time. Under the microscope, the green-colored microtubule ropes tagged with GFP could be seen tugging on the X-shaped chromosomes to tear them apart and place each half into one of the newly made daughter cells.[33] But Prasher's role in the discovery was not forgotten by his colleagues. Both Chalfie and Tsien paid for Prasher to attend the Nobel Prize ceremony in Sweden, where all three recipients thanked him in their speeches. After a long hiatus from scientific research, Prasher returned to the lab in 2015 to work with Roger Tsien at University of California in San Diego.

LIGHTING A WAY FORWARD

The emergence of fluorescence in biology represents yet another way in which animals have developed ways to produce light. It is still not exactly clear why jellyfish use GFP for biofluorescence, but it's possibly a way to ward off predators. Against the blackness of the ocean, the trailing flashes of light scattered across their expansive network of tentacles may confuse predators, making it difficult to figure out where the jellyfish begins and where it ends. To us, human bystanders, it is a mesmerizing light show and another fascinating example of how proteins produce the most unusual features in some of the most mysterious creatures. Yet we are not merely enchanted bystanders. In fact, scientists have used the illuminating properties of GFP to make some of the most important discoveries in biology.

In the decades since the discovery of GFP, scientists have modified the physical properties of the protein by making changes to its individual building blocks. GFP, as it turns out, is quite malleable, and small mutations in its amino-acid sequence can dramatically change the way it behaves toward light. To date, numerous GFP variants have been engineered, each exhibiting fluorescence at a different color.[34] With such a wide selection of engineered fluorescent proteins, from yellow fluorescent protein and cyan to red, blue, and citrine, scientists can now track the location and movement of more than one component within the same living cell by tagging each component with a protein that gives off a different color of light. Together, the glow of different colors under the microscope paints a canvas of molecular architecture and illuminates a microscopic world that was elusive not so long ago. With the rainbow of colors produced by modifying the jellyfish GFP, scientists have also been able to map the intricate connections of millions of neurons in the brain of a mouse. As each cell was given instructions to produce a unique combination of different fluorescent proteins, the entire brain lit up with hundreds of color combinations, each color unique to one cell type. This "brainbow" elegantly displayed the intricate topology of the entire brain in never-before-seen detail.[35]

Not only are fluorescent proteins used to map the brain, but luciferases, like those found in the firefly or the click beetle, are used in molecular imaging, too. More sensitive than most fluorescent proteins, luciferases can track the ebb and flow of many biological processes. One major application for luciferases has been cancer research.[36] Tumors growing in laboratory animals are often tagged with one of the many luciferase enzymes found in nature. By monitoring the amount of emitted light, the growth or spread of the tumor can be precisely monitored. Moreover, this strategy can help scientists monitor how a tumor responds to an experimental drug in real time, without having to extract the tumor from the animal. Fluorescence and bioluminescence have also been deployed in the fight against pollution. Recently, *E. coli* bacteria were modified to produce a version of luciferase only when challenged by environmental stresses.[37] The bioluminescent signals that the bacteria produce have been used to detect cancer-causing pollutants, DNA-damaging molecules, and toxic heavy metals like arsenic or antimony.

From the mysterious light displays put on by ocean animals, to the enchanting glow of fireflies, luminous creatures have always inspired awe and wonder. These feelings have only been enhanced by the discovery of the proteins involved, and our growing understanding of how these proteins have contributed to these organisms' survival. As we look ahead to new scientific breakthroughs, there is no doubt that fluorescent and bioluminescent proteins will play a crucial role in guiding the way. The gentle glow of a firefly on a warm Indiana summer night reminds us of the creativity of nature, and points to the creativity of the human mind—our ability to harness the most brilliant tools from nature to decipher the mysteries of the world and develop the technologies that will shape our shared future.

7

REMEMBERING

SHAHIR S. RIZK

I CAN'T IMAGINE WAKING UP each morning having to relearn what my name is, what I do for a living, where my work is, or, God forbid, how to work the coffee machine. Most animals can store information to use later; even a goldfish can remember which side of the tank the food comes from. Our pets, especially dogs, remember our scents for life, leading to stories of reunions between dogs and their people even after years of separation. A particularly impressive example of remembering is how salmon return after years of roaming the open ocean to the same stream where they were hatched.

Although memory clearly is crucial to the survival of animals, it may be more surprising to learn that it is also vital in many other species, including plants. I remember the first time I saw a Venus flytrap in the wild. I was hiking with my wife along a trail at Carolina Beach State Park in North Carolina. The air was thick with humidity and the salty scent of the ocean. "I thought they would be bigger," I told my wife, peering down at the tiny insect-eating plants along the sides of the trail, fragile and barely noticeable. I had pictured them as great monsters that would clamp down on my hand and break off my thumb if I didn't get away fast enough. Yet their prey—gnats, ants, and fruit flies—are typically much smaller than any of my fingers. The flytrap has jaw-like leaves lined with "teeth," which interlock when the hinged leaves close down on an insect. Growing along the sandy nutrient-poor soil, the Venus flytrap cannot get enough nitrogen, an essential element for making DNA and proteins, through its root system. So instead, the plant gets nitrogen the way most animals do, by eating some other organism. The prey, lured into the flytrap's "jaws" by a fragrant nectar, triggers the leaves to snap shut, trapping itself inside. Within the closed jaws, digestive enzymes secreted by the plant, as well as the acidic environ-

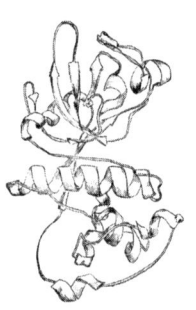

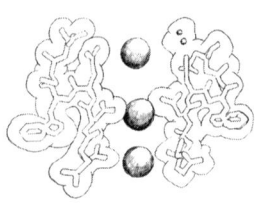

ment, slowly break down the prey into valuable nutrition.

How the leaves snap shut had puzzled scientists for centuries. After all, we don't typically think of plants as being able to move very quickly, let alone eat an animal. The carnivorous nature of this plant was not established until well into the eighteenth century.[1] Even Carl Linnaeus, the father of taxonomy (the science of classifying organisms), did not believe that a plant could hunt an animal, declaring that it is "against the order of nature as willed by God." A few decades later, Charles Darwin fell in love with the Venus flytrap, performing many experiments on it. He called the plant "one of the most wonderful in the world" and wrote a whole book about carnivorous plants.[2] It is only relatively recently, however, that the mechanism by which the leaves clamp down on their prey was discovered.

When an insect is captured by the Venus flytrap, many enzymes are released and begin to digest the prey: one of these enzymes, the Venus flytrap protease (PDB code 5A24), breaks down the proteins of the captured insect. The mechanism by which the Venus flytrap snaps shut is regulated by the flow of calcium ions. Shown here is a closeup of a calcium channel protein (PDB code 5KLB), with the calcium ions represented as spheres passing through the narrow channel.

To trigger the trap, tiny hairs along the leaves of the plant must be touched by the prey. Touching one hair is not enough; two hairs on the same trap must be touched, one after the other, within seconds. This means that the plant must be able to "remember" the first touch and anticipate a second touch within a certain time period.[3] If the second touch comes too late, the plant "forgets" the first touch ever happened. The strange way in which the plant responds to its environment and stores that information in short-term memory is facilitated by a unique set of proteins, most notably a gatekeeper protein known as a calcium channel. Like a double door, the calcium channel sits between two compartments in the plant cell. It is normally closed, preventing calcium from moving from one cellular compartment to another. But when one hair is touched by a potential prey, the calcium channel opens, and some calcium is released. The amount of calcium released is ultimately what triggers the trap to close on the prey. One calcium release event is not enough to trigger the trap. Touching another hair triggers the release of more calcium. Only then is there enough calcium to activate the clamp to snap shut in one swift motion.[4] If the second touch comes too late, the calcium released from the first trigger is slowly mopped back up, and the plant begins to forget. In a very simple and elegant way, the Venus flytrap uses proteins that move calcium between two compartments to establish a memory of something that it experiences. It may not be the same kind of memory that we have, but nonetheless, it is a way to remember. It is also a way to forget.

TINY, MUTABLE MEMORIES

Outside the old research building named after Nanaline Duke, the second wife of the founder of Duke University, a group of biochemistry graduate students would gather around a trash can, smoking cigarettes and sharing stories of failed experiments. Some listened, some talked too much, but they all took solace in knowing that there were others who could relate to their problems. Scientists, it seems, speak a different language. Your therapist might help you get through losing a pet or a failed relationship, but a therapist may never understand the pain of getting results that don't make sense or losing a protein sample you worked months to obtain. Each of the graduate students knew the effects of smoking. Some even studied lung cancer. Yet those few moments shared with other graduate students always seemed worth it.

A puff of nicotine-laden smoke and the company of like-minded friends going through the same kind of suffering eased the pain and frustration of all those failed experiments.

Nicotine is not only one of the most addictive substances we know of; many of the components in tobacco smoke can also trigger mutations in our DNA that affect the growth and proliferation of cells.[5] Often, mutated cells become "immortal," dividing endlessly and forming cancerous tumors that spread to other parts of the body, eventually leading to death if not detected early or treated. Yet cigarette smoke is only one of thousands of carcinogens that humans are exposed to. Every day, our bodies are bombarded with substances that can alter the sequence of our DNA. Even sunlight can cause mutations with its powerful ultraviolet rays.[6] Mutations can also arise spontaneously, and often do, especially during cell division, the time when a cell must make a copy of its DNA for each of the newly produced cells.[7] Mistakes in copying the exact sequence of the original DNA can result in mutations in protein sequences, so the process must be tightly controlled.

The main protein responsible for copying DNA is the amazing enzyme DNA polymerase. As we explained in Chapter 5, DNA polymerase "reads" the genetic code, made up of A, C, G, and T bases, and makes a matching copy. It does this by sliding along the "parent" DNA in a single direction like a ratchet and making a new "daughter" DNA strand. The new DNA strand is not identical but is complementary. For example, when DNA polymerase encounters an A on the parent DNA, it adds a T to the growing daughter DNA strand because A always pairs with (complements) T. When DNA polymerase encounters a C, it adds a G to the daughter strand because C always pairs with G. DNA polymerase does a fantastic job matching the sequence of the new DNA with the parent DNA, but it is not perfect; every now and then, it makes a mistake. Its error rate is about one in a hundred thousand letters.[8] That may still seem impressive (and it is), but considering that a human cell has roughly three billion DNA bases that must be copied each time the cell divides, this amounts to about thirty thousand mistakes in each DNA strand. Each mistake is a potentially deadly mutation. Even our favorite lab bacterium, *E. coli*, contains roughly 4.6 million DNA bases in its genome, which amounts to nearly fifty errors each time the cell divides. Considering that *E. coli* can double in number every twenty minutes, these mistakes can add up quickly and cause catastrophic outcomes.[9]

Around the trash can with the ashtray on top, I stood among the rest of the graduate students, lamenting my failed experiments and looking for any advice and emotional support I could get from the others. Every now and then, a man with a perfectly groomed mustache would walk by and shoot the breeze with us. We all knew him. His name was Paul Modrich, and he taught our nucleic acids class. Modrich had a modest lab on the first floor of Nanaline Duke. He was interested in DNA damage and how DNA was copied with such high accuracy during cell division, even with the high number of mistakes made by DNA polymerase. When he began working on this problem, there was already evidence of molecular machinery dedicated to repairing DNA replication errors, an entire set of proteins and enzymes capable of identifying the mistakes and correcting them. But no one had identified those proteins yet or figured out how they worked.

Over several decades, Modrich's lab helped isolate the proteins, which belong to the "Mut" family (pronounced "mute"). Through a set of clever experiments, they showed that these proteins follow in the footsteps of DNA polymerase and fix any mistakes they find.[10] For instance, if Mut proteins see an A paired with a C or a G instead of a T, they know that a mistake has been made and they go to work fixing it. Without Mut proteins keeping watch over the constant

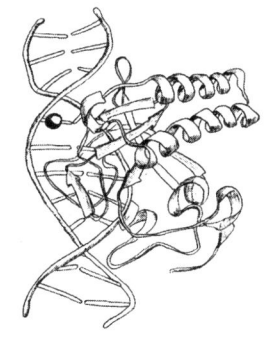

Many proteins are involved in maintaining the fidelity of DNA replication, the process of copying DNA during cell division. One of the main proteins that checks for potential mistakes is MutH (PDB code 2AOR). This protein binds to DNA and looks for errors in the newly made DNA strand. It identifies the old strand by finding a methyl group, shown here as a sphere. This allows it to spot the correct sequence found on the old strand and recruit other proteins to repair any potential mistakes.

emergence of mistakes, bacteria, which divide rapidly, would lose all sense of what to do within just a few hours. In fact, Mut proteins were first discovered in bacteria, and this is how they got their name; the term Mut is short for "mutator." When scientists found bacteria that rapidly accumulated mutations, they found that those bacteria had a set of dysfunctional proteins that they called "mutator" proteins. Later, they found that these so-called mutator or Mut proteins in fact fix mutations as they arise. It was their absence that made the bacteria so prone to mutations.

To understand how Mut proteins fix mistakes, Modrich and his team took advantage of a set of proteins known as restriction enzymes. Each restriction enzyme breaks the DNA double helix (double-stranded DNA) at a precise site. Even a difference of one base can prevent the action of the enzyme, and the DNA can't be cleaved. The researchers began with a double-stranded DNA segment, with one strand containing the correct sequence for cleavage by the restriction enzyme, while the other contained a single mutation. That one difference resulted in a mismatch between the two DNA strands of the double helix. Only when the mismatch was repaired could the restriction enzyme break the DNA segment into two pieces. By monitoring the action of the restriction enzyme, they were able to determine the role of each of the Mut proteins in the repair process and understand the conditions under which they function.

But as clever as Mut proteins are, they are faced with a problem: how do they know which is the correct base? For example, if they encounter an A paired with a C, the question becomes: is the A the correct sequence (and there should be a T on the other side), or is the C the correct base (and there should be a G in place of the A)? This would be an incredibly difficult problem if the cell had no way of remembering which is the parent strand—the one with the correct sequence—and which is the daughter strand—the newly made one likely to contain the mutation. Fortunately, our cells have a simple and very effective way to remember which strand is original and which is new.

We typically know how old things are by how much wear and tear they have accumulated. My favorite pair of jeans have some marks that blend in with the denim pattern, along with tears, scuffs, grass stains, and frays. In a similar manner, the older DNA in bacteria accumulates modifications the longer it's been around in the cell. In bacteria, special proteins mark DNA by adding a chemical group known as a methyl. This tiny chemical modification, amounting to a single carbon surrounded by three hydrogens, is added by a protein known as DNA methylase, which scans along the DNA until it finds a specific sequence, GATC, and adds a methyl group to the A. Once Mut proteins find a mismatch in the DNA, they quickly scan both strands to see which one contains the methyl group on the A at the nearest GATC sequence. This tells them which is the parent strand containing the correct sequence, and which is the daughter strand containing the mistake left by the speedy, but not entirely accurate, DNA polymerase. One protein, called MutH, latches onto the daughter strand, the one without the methyl modification, and recruits other Mut proteins. Once signaled by MutH, the other Mut proteins get to work replacing a small stretch of the DNA double helix that contains the mutation with one containing the correct sequence.

Without the work of Mut proteins, DNA sequences would fall victim to too many mutations, many of which would lead to catastrophic outcomes. In all known organisms, DNA sequences act as the hard drive of cellular information, providing the genetic memories that get passed down from generation to generation. Without tight control of the fidelity of DNA replication, all these memories would be distorted. Working tirelessly to fix mistakes when DNA is copied, Mut proteins are essential to the way we pass down genetic information from one generation to another, and how our cells prevent disease-causing mutations from happening. The significance of Mut proteins has been recognized by the scientific community. In 2015, Paul Modrich was awarded the Nobel Prize for his work on how cells maintain their genetic memories, a recognition of decades of work deciphering the function of our DNA repair machinery.[11]

MOLECULAR MEMORIES

My first winter in America was one of the coldest on record. The temperatures outside our little apartment dipped to $-30°F$ ($-34°C$), while snowstorms raged as far south as Florida. In January, I sat in bed, weighed down by a fever and headaches. "It's a virus," my dad said, "Nothing to worry about—new place, new bugs." And it was in fact a new bug, nothing my immune system had seen before, and so I lay helpless until my body could fight it off. At least I got a day off from school, one less day of teenage awkwardness, one less day of translating my thoughts into strange English, one less day of having to explain that I didn't own a camel and I didn't live in a pyramid.

While I lay in bed, in and out of fever dreams, my immune system was working in overdrive. When the virus found its way into my body, a whole set of physiological responses was set in motion. The fever that put me in bed was nothing more than an inflammatory response. Meanwhile, a special group of immune cells was busy at work. B lymphocytes are a type of white blood cell that springs into action when an infection takes place. When life is good, these cells lay dormant yet on guard, circulating around the body, looking for intruders. On the surface of each B cell is a unique protein receptor that juts out in the shape of a Y. Its two branches have precise and complex structures at the very ends that resemble hands or cavities. Since our immune system contains about 10 billion different B cells, each with a unique B-cell receptor shape, the variety of cavities can fit an immense number of shapes, corresponding to the structures of proteins on the surface of invading bacteria or viruses.[12] When the "hands" on the B-cell receptor bind to a protein that they suspect is from an invader, they begin the process of neutralizing the threat. But first, the B cell must make absolutely certain that the protein it grabbed hold of is in fact a foreign protein and not one that belongs to our own body. To do this, a B cell pulls the possible foreign protein inside and digests it into smaller pieces, which are then presented on the surface for another kind of immune cell, known as a T lymphocyte (or T cell), to check out.[13] The T cell is an expert at distinguishing the difference between our own proteins and those that belong to invaders. Once the small chunks of a potential threat are displayed on the surface of B cells, T cells grab onto the suspects, probing and investigating their shape. After close inspection, if the T cells determine that the protein in question does not match any of our own, they secrete signals in the form of cytokines: proteins that trigger B cells to take decisive action against the invader. Cytokines can have a profound effect, activating a remarkable transformation in the shape and function of the B cell.

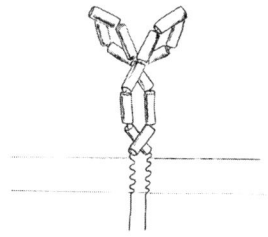

Activated B cells, triggered by T-cell cytokines, begin to grow, divide rapidly, and transform into what are known as "plasma cells." The B-cell receptor on the surface—the Y-shaped protein that initially recognized the foreign protein—becomes an antibody, a foot soldier in the fight against an infection.[14] The activated plasma cells become antibody-making factories, secreting billions of copies of antibodies into circulation, which then seek out the invading bacteria or virus. Once released, the antibodies fight the invaders in numerous ways. First, they can surround the bacteria or virus by binding all over its outer shell, preventing it from invading our own cells and slowing down its growth. Second, the antibodies' "arms" can grab onto two invaders at the same time, and many antibodies together can join hands like a daisy chain to bring multiple invaders together in one "clump." This is a process known as agglutination, whereby antibodies concentrate invading organisms into clusters, making it difficult for them to move and providing a large target for the rest of our immune cells to attack. Finally, like beacons, antibodies can "tag" the proteins on the surface of an invader and recruit a special type of immune cell known as a phagocyte.[15]

B cells are a major component of our adaptive immune system; their primary function is to produce antibodies to fight infections. Before antibodies are secreted, they begin their life as B-cell receptors with a characteristic Y shape. They reside on the surface of B cells and try to catch foreign invaders. The two horizontal lines represent the cell membrane of the B cell from which the receptor points toward the outside of the cell.

This cell has an appetite for germs. Its name literally means "devouring cell." A phagocyte will seek out invaders surrounded by the antibody beacons and swallow them whole, then break them up into smaller pieces with the help of a soup of acidic digestive enzymes.

B cells are essential for what is known as adaptive immunity: our body's way of responding to the ever-evolving world of germs and invaders. When triggered by a new bug, B cells spring into action, turning into plasma cells and sending their molecular soldiers—the Y-shaped antibody proteins—to sabotage the invader's attack. Once the threat is neutralized, plasma cells slow their production of antibodies and preserve their resources as life goes back to normal. Yet the memory of an infection often becomes etched deep into our immune system.[16] When B cells are activated, they don't all turn into plasma cells. Some morph into memory cells and make their way into the lymphatic system, a network of tubes that white blood cells use to travel throughout the body. There they lay dormant, waiting for the invader to return. Memory cells are our immune system's way of remembering the precise shape of a foreign protein, one carried by a pathogen with bad intentions. By preserving a memory of the shape of the invader from an earlier infection, the memory cells remain on guard, ready to launch a quicker attack if the invader ever returns. While the initial activation of a B cell into a plasma cell can take up to two weeks, the reactivation of a memory B cell takes only two to four days, guaranteeing that the quickest defense system possible is at the ready.

By February, I was feeling stronger, and the memory of the virus was long gone from my thoughts. The imprints of its proteins, however, were still carried in my bones. By March, I was cold-hardened. I endured the rest of winter as if I had lived in South Dakota my whole life. I got used to the sun gliding across the horizon, pale and lonely, illuminating the endless silver fields of snow. I was getting taller, spoke faster, and had learned a lot about pep rallies, American football, church potlucks, and Chinese open buffets. I didn't mind walking to school in the chilly wind or the girls asking if I had a girlfriend back in Egypt. I didn't even mind ten inches of fresh snowfall in April, or the green-gray clouds of the tornado that grazed the town two weeks later. And all these new experiences became memories that I can access today.

SCARRED MEMORIES

Long before I lay in bed with a new virus, a much older virus took hold of Europe. For centuries, outbreaks of smallpox swept through cities and the countryside, wiping out as many as 30 percent of those afflicted.[17] Even those who survived carried with them visible scars on their faces and bodies—a fate, for many, that was nearly as bad as death, because they were marked forever by disfigurement and often shunned by their communities. In the early 1700s, Lady Mary Wortley Montagu, an educated woman of nobility, left England for Constantinople with her husband, who had been appointed ambassador to the Ottoman Empire.[18] During her time in Constantinople, she noticed how smallpox was much less deadly there and became determined to find out why the Turkish people were more resistant to it. She was motivated not only by her curiosity, but also by personal experience; only a few years earlier, her brother had died from the disease, and she herself had barely survived a battle with the illness, which had robbed her of her beauty and left her with permanent scars. With a young son and a newborn daughter, Lady Mary was desperate to protect her children from the terrible plague. The noblewoman befriended many Turkish women, who eventually invited her to a special ceremony carried out to protect children from smallpox. During the ceremony, the women took a small amount of pus from a patient with smallpox and rubbed it into a small scratch on the skin of a healthy person. The person would inevitably develop a fever, then recover, having immunity to the disease. This process, known as inoculation, had been carried out for centuries in parts of Asia and Africa but was unknown to Europeans until Lady Mary brought it back to England. She had her son inoculated but feared that her daughter might be too young for the procedure.

The key to the success of inoculation is how it introduces the virus to the body. Usually smallpox is transmitted through the air and colonizes the tissue of the lungs and the respiratory tract.[19] That is when it's most deadly. But introducing the virus particles directly into the bloodstream proved to be a safer way to form immunity. B cells are quick to fight the infection before the virus reaches the lungs, where it usually does the most damage. Yet inoculation is not without its dangers. Introducing too much of the virus could have devastating effects, and introducing too little may not trigger any immune response. Around 2 percent of people inoculated with smallpox died. But those who survived were forever immune to the illness, carrying only a small scar on their skin where the virus had been introduced.

By 1721, Lady Mary and her family had returned to England, where another outbreak of smallpox was raging through the country. While most people were forced to quarantine, Lady Mary saw an opportunity to spread the word about inoculation. She enlisted the help of a local doctor, James Keith, who had lost two of his sons to the disease.[20] Skeptical at first, Keith eventually tried the procedure on one of his children after trying it first on Lady Mary's daughter. The children experienced mild fevers and then recovered, suffering no severe symptoms and developing no scars. Convinced of the procedure's success, Keith worked with Lady Mary to launch an inoculation campaign throughout London.

The same year of the outbreak in England, a similar outbreak took hold across the Atlantic, in the new colony of Boston.[21] At the time, a prominent minister named Cotton Mather was concerned about the spread of the disease and was looking for ways to stop it. Mather, who notoriously participated in the Salem witch trials, sought the help of an enslaved man who had been purchased by Mather's congregation. The West African man, named Onesimus, spoke of a procedure he had undergone as a child, which had protected him from smallpox for the rest of his life.[22] The procedure he described was similar to the inoculation witnessed by Lady Mary during her time in Constantinople. While Mather did not fully trust Onesimus, he was intrigued by the procedure and desperate enough to try it, having seen so many members of his community die from the disease.

Using Onesimus's experience with inoculation, Mather began to try the procedure on willing Bostonians with the help of a local physician, Zabdiel Boylston. While highly effective, the procedure was not completely harmless; of the 242 people inoculated, six died, or one in forty.[23] But that was still a much better outcome than getting the disease through normal means, which could come with as much as a one in three chance of dying. Even though inoculation was successful in slowing the smallpox outbreak, the process was met with resistance from the community. While some were skeptical of a technique brought about by a Black man, others saw it as witchcraft. The success of inoculation, as evident by the number of survivors, was ignored, and a strong resistance mounted against Mather, Onesimus, and Boylston. Newspapers attacked the men, and at one point an explosive package was thrown through Mather's window.

Back in England, a similar type of resistance was mounting against Lady Mary's efforts to inoculate her community. Xenophobia was rampant, and most Europeans were suspicious of medical techniques developed in the Middle East, Africa, or Asia, where people were considered uncivilized. Yet Lady Mary's efforts did not stop. With the help of Keith, she conducted inoculations on any willing participant. They also carried out inoculations on orphans and convicted criminals, who were viewed as less valuable, a practice that would be highly unethical by today's standards. Even when Lady Mary convinced the Princess of Wales to inoculate her own daughter, resistance to the technique persisted among the public. Some saw it as a political move and refused the procedure on those grounds alone. Lady Mary spent the remaining years of her life promoting the practice of inoculation. She wrote that her efforts have been "arduous, fearful, and thankless." Yet thanks to her work, thousands of adults and children were protected from the virus. One of those children, a boy named Edward Jenner, would grow up to become known as the father of vaccines.[24]

As an adult, Jenner learned that milkmaids often got a milder form of the smallpox disease, a version commonly known as cowpox. Cowpox is usually transmitted when a person comes in contact with the pus from a lesion on the udder of a cow. The first symptoms, like those of smallpox, included fever and blisters on the hands and face. But cowpox patients almost always recovered quickly and completely, with no scars. Jenner postulated that the two viruses might be similar enough that getting immunity against cowpox could prevent serious infection from smallpox. In 1796, he conducted a famous experiment, also highly unethical by today's standards. Jenner inoculated a boy with cowpox, and when his symptoms subsided, he challenged the boy's immune system with smallpox collected from infected patients. The boy appeared immune to infection, and the term "vaccination" was coined, based on the Latin word *vacca,* meaning cow. Vaccination involved challenging the immune system with a weaker version of the virus—in Jenner's case the cowpox virus—to build immunity against a severe disease.

Like the discoveries by Mather and Lady Mary, Jenner's findings were met with skepticism. Many of the same attacks against inoculation were used to discredit vaccination. Some clergy claimed it was against God's will to introduce materials from an animal into man. Newspapers even published cartoons showing vaccinated individuals turning into cows.[25] Sadly, many of the same attacks used during the early days of vaccination are still common today among those who refuse immunization. Superstitions and misconceptions that were endemic in the so-called pre-Enlightenment European mindset remain rampant in our own supposed Age of Reason.

Despite resistance from anti-vaccine activists, the smallpox vaccine continued to be developed through the twentieth century, time and time again showing its effectiveness at preventing infection and death from the devastating disease. The vaccine continued to use the virus derived from cowpox to give people's immune systems a lasting memory of the far deadlier smallpox virus. In 1980, the World Health Organization declared smallpox eradicated, a testament to the resounding success of immunization and the efforts of marginalized and often forgotten figures like an enslaved West African man and a disfigured English woman.[26] Having been born in Egypt just two years before smallpox was declared eradicated, I became one of the last people to be immunized against the disease.

While I never saw a person with smallpox, I did see children with polio in my school. They suffered paralysis in their lower bodies, wore braces around their legs, and moved around with difficulty on crutches. While some of the mean kids bullied them, the rest of us pitied them and were grateful to have fully functioning legs. Nowadays, there are almost no cases of polio in the United States, or in most countries.[27] Those who used to die or live with disabilities from the disease are almost completely forgotten. With no major plagues in living memory, to some it may not seem necessary to get immunized. After all, what are we getting immunized against if there is nothing killing us?

Millions of lives have been saved and continue to be saved through immunization. As a result, life expectancy worldwide has increased,

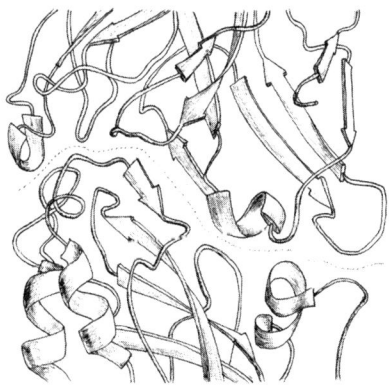

A closeup of the interaction between an antibody and a protein from *vaccinia*, a virus used in the smallpox vaccine (PDB code 6B9J). The antibody (*top*) uses exquisite shape complementarity to match the surface structure of the pox protein (*bottom*), ultimately blocking its function and targeting it for destruction. A dashed line indicates the contour of the two interacting surfaces.

leading to unprecedented population growth. My own grandparents, who married in the late 1930s, had a total of twelve children. Only six survived. Of the nine children born before 1950, before vaccines and antibiotics were widely available, only three survived. By contrast, all three of their children born after 1950 survived into adulthood. My grandmother spoke fondly of the children she lost, most of whom died before the age of one. The way she spoke about their deaths was a reminder that this kind of tragic survival rate was commonplace not so long ago. Only a couple of generations later, we have already forgotten how the odds of survival were stacked against us until vaccination tipped them in our favor.

Immunization is an exercise in molecular memory. A flashcard shown to our immune cells in the form of a weakened microbe or a protein from a pathogen puts the immune system on high alert. It marks the enemy, leaving a memory carried by the aptly named memory cells that roam the body for years or decades, waiting for the invader to return. Yet our immune system's memory is not always perfect. Some vaccines, like those for the smallpox or polio viruses, can be highly effective and build a nearly lifelong resistance to the disease. Other vaccines are not as effective. For example, people living in areas where diseases like malaria are endemic can build some form of immunity.[28] But that immunity tends to fade if they move to places where they are not continually exposed to the disease-causing parasite. With many viruses, the problem of waning immunity happens for a different reason. Every year, we get the flu vaccine, but that's not because our immune system forgets. It's because the virus mutates, changing its identity ever so slightly to evade our memories of it.[29]

The process by which our bodies remember an invader protein is more complex than anything I can write about in these few pages. In fact, much is still unknown about our immune system, and how each component works together to fight off disease. For example, it is still not entirely clear why some invader protein signatures are more memorable than others and how, in some cases, our immunity can wane so much that we fail to remember a past infection. What we actively remember, what we carry along in a dormant state for later, and what fades into oblivion are hard to predict.

HIJACKED MEMORIES

For a long time, it was thought that only sophisticated organisms like humans or other vertebrates could exhibit characteristics of an adaptive immune system, one that can remember past infections. This is because of how complex the process of keeping a memory of an infection can be. After all, it requires multiple types of cells to sort through which proteins belong to an invader and which proteins belong to us. The idea that a single-celled organism could remember a past infection seemed far-fetched once. But now it is clear that even bacteria can remember.

In the wild, the major infectious agent of bacteria is a class of viruses known as bacteriophages. These viruses, while harmless to us, use bacteria to make more copies of themselves because they lack the protein machinery necessary for their own reproduction.[30] When a bacteriophage (or phage for short) invades a bacterial cell, it injects its genetic material and hijacks the cellular enzymes for its own purposes. Immediately, bacterial proteins, like DNA polymerase, begin to make more copies of the phage DNA, and the bacterial cell essentially becomes a phage factory. When the new phage particles are released, they go on to infect more and more bacteria, and the chain reaction continues.

Occasionally, some bacteria manage to survive the infection, and in the process, they take a piece of the phage DNA and store it in their own genome.[31] This renders the bacteria permanently changed, scarred from battle, but able to pass along that piece of DNA to future generations of bacteria. Surviving a phage infection leaves a memory collected in the form of a piece of DNA, stored in a special region within the bacterial genome. The region where these memories are kept was discovered by accident when scientists found a string of phage DNA sequences interspaced within the bacterial genome. The scientists named this region the "clustered regularly interspaced short palindromic repeats," or CRISPR for short.[32]

In a way, CRISPR is a type of adaptive immune system, where memories of past infections are stored. When a bacterium is challenged by a foreign phage, the DNA from the invader is compared to the memories stored in the CRISPR region. If a match is found, then a special protein known as Cas9 springs into action. Cas9 is an enzyme that becomes activated in response to infections and uses the CRISPR sequence as a guide to seek and destroy the invader's DNA. This simple mechanism may be a far cry from our own complex, adaptive immune response, with all of its intricate steps, but it is nevertheless highly effective. To date, the CRISPR response has been found in about 40 percent of bacteria and nearly 90 percent of archaea, bacteria's more ancient cousins.[33]

When the mysterious regions of CRISPR were first discovered in the 1980s, scientists had no idea what they meant. It wasn't until the early 2000s that a connection was made between these sequences and the sequences found in the genomes of bacteriophages. When the Cas9 enzyme was initially identified, it was thought to play a role in DNA repair because of its ability to cut DNA at very precise sequences. Eventually, Cas9 was shown to cut the DNA of invading bacteriophages, rendering them ineffective. Scientists soon realized that Cas9, with its ability to target DNA for destruction, could be a useful biotechnology tool. The ability to cut and paste DNA at precise sequences became an important implement in modifying DNA sequences not only in a test tube, but also within living organisms. In an ironic twist, the Cas9 enzyme meant to preserve memories of past infections could now be used to delete harmful mutations in other organisms, effectively erasing memories of inherited diseases. With some modification, the CRISPR / Cas9 system has been used in gene editing to cut and paste genetic material in a variety of ways. Recently, the system has been used, with promising results, to repair inherited genetic mutations in human diseases. For instance, it has been used to treat sickle cell anemia, as well as leukemia and other blood cancers, by modifying a patient's own immune cells so that they attack and destroy cancerous or otherwise malfunctioning cells.[34] In a sense, we are using an ancient, relatively simple bacterial immune system to help boost our own.

When Cotton Mather asked Onesimus if he had had smallpox, the enslaved man's answer was somewhat confusing. Onesimus's response was both yes and no.[35] He had been exposed to the deadly virus in order to survive future attacks. Onesimus showed Mather the scars on his arm from inoculation, a reminder of how the virus is always with him but also not. Today, I carry my own scars of viral memory—one from the smallpox vaccine on my thigh and two from the tuberculosis vaccine on my shoulder. I am carrying memories of these invaders deep in my molecular composition, the way our ancient relatives, bacteria, have done for eons, since they first began to gather pieces of viruses in their genomes.

8

DEFIANCE

SHAHIR S. RIZK

IT WAS THE MIDDLE of winter, but the sun beat down mercilessly on my forehead. As I looked over the bridge, I could see the large lotus flower columns flanking it on either side. In front of me was the Nile River flowing north, its banks mottled with tiny patches of green palm trees. Behind me was Lake Nasser, a massive body of water that was for decades the largest manmade lake in the world. I was standing on the High Dam, a massive concrete structure built in the mid-twentieth century at the southern border of Egypt to regulate the flow of the river and to provide hydroelectric power for a steadily growing population. The dam sits just south of the majestic city of Aswan, splitting water between river and lake. Turning my head to the left, I looked west into a scene that contrasted with the lush patches of green along the Nile banks. I was staring into the abyss of the Sahara Desert, the largest hot desert in the world. The image of a sea of sand dunes stretching as far as I could see would forever be carved in my mind. Knowing that this vast desolate ocean of sand and stone stretched across the widest part of the entire continent of Africa gave me an immediate sense of terror. The juxtaposition of wet and arid, green and brown, life and death is hard to miss when you are staring at the vastness of this landscape from the unique vantage point of the dam.

Nearly ten thousand years ago, this desert was teeming with life, a vast savannah that was home to large mammals—gazelles, lions, giraffes, and buffaloes with herds numbering in the thousands. It was lush with trees and shrubs, feeding countless mammals and housing an immense number of birds. Large crocodiles swam in rivers and ponds filled with hundreds of species of aquatic animals. But a shift in the Earth's axis and a decline in rainfall forever changed the landscape, creating the emptiness that spans an entire continent.

Yet even with the loss of rainfall and the scorching heat, life thrives in the most unusual ways. Hidden from the sun, under rocks and inside shady canyons, lies a creature whose image is as fear-inducing as the image of the terrifying desert itself. With eight segmented legs and a long, curved tail, the deathstalker scorpion is the stuff of nightmares. As one of about two dozen species of scorpions native to Egypt, the deathstalker has been part of the culture for millennia.[1] In many ways, the story of the scorpion is Egypt's origin story, dating back to its earliest dynasties. It is said that the first king of ancient Egypt was a scorpion. Maybe not an actual scorpion, but a man who took the name of the animal as his own. Not much is known about King Scorpion I or his successor, King Scorpion II. But some say that Scorpion II was the same man who united northern and southern Egypt, creating an empire that lasted more than five thousand years. The symbol of the scorpion also pervaded the mythology of ancient Egypt. The goddess of healing, Serket, considered the protector from snakes and poisons, carried a scorpion on her head.

The people of the city of Aswan have coexisted with the scorpions for as long as people have inhabited the Nile Valley. Most of the time, the scorpions keep to themselves, but when threatened, they will sting. In 2021, a heavy rainstorm, unusual to this part of the world, caused large-scale flash flooding, bringing out thousands of scorpions from their hiding spots and driving them into the city. The scorpions sought new hiding spots, like shoes and sandals, stinging more than five hundred people.[2]

While not always deadly, a scorpion sting is very painful. With a quick strike of its tail, the deathstalker scorpion delivers venom loaded with several toxins, including a protein known as chlorotoxin.[3] This short protein targets the nervous system, causing paralysis and intense pain. It helps the scorpion immobilize its prey, making it easier to devour or ward off large predators by disrupting nerve function. To transmit signals from the brain to our extremities, nerves rely on ion channels. These tiny double doors sit on the membrane of nerve cells and regulate the flow of ions. The flow of ions and their charges across

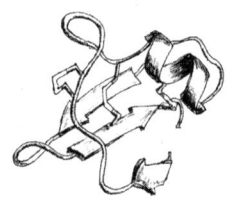

a membrane triggers the electrical signal that moves through our nerves at near lightning speed, facilitating everything from movement to digestion to the secretion of hormones. After the scorpion injects it into a victim, chlorotoxin finds and latches onto the specific type of ion channels that regulates the flow of negatively charged chloride ions. It quickly stops the ion flow, interfering with normal nerve function and signal transmission and causing paralysis and severe pain. The deathstalker scorpion can go for weeks without a meal, but armed with chlorotoxin, it can immobilize prey larger than itself within seconds. This tiny protein gives the scorpion a clear survival advantage in the harsh and unforgiving landscape. Proteins have not only evolved to be the building blocks of living things and catalysts for life—they have also become instruments of death.

Chlorotoxin is one of the proteins found in the venom of many scorpions. The protein blocks the flow of chloride ions, causing paralysis and intense pain in its victims.

Throughout all the kingdoms of life, we find examples of powerful proteins used not only for self-defense but also to attack. Across the animal kingdom, nothing creates a more terrifying image than that of a poisonous snake with a scaly slithering body and a set of sharp fangs ready to strike and deliver a deadly venom. The venom of most snakes is a cocktail of proteins, each of which can kill in a different way. The spitting cobra is truly horrifying. Met with a predator or a perceived threat, the cobra will shoot a blinding venom containing dozens of protein toxins. Every year, more than five million people are bitten by snakes, and roughly a hundred thousand of them die.[4] Cobra bites can have up to a 20 percent death rate.[5] One of the proteins in the cobra venom is cobratoxin. It belongs to a family of proteins known as three-finger toxins.[6] Their name comes from their structure, which contains three protrusions made up of beta strands and wiggling loops.

Cobratoxin is a powerful neurotoxin. Even a small amount can take down large prey. The key to the toxicity of co-bratoxin is how it interferes with the signals between nerves and muscles. Normally, nerves transmit chemical signals from the brain, in-structing a muscle to contract. These signals are mediated by a small molecule known as acetylcholine, a tiny chemical neurotransmitter (a class of chemicals that act as the signal be-tween nerves or between nerves and muscles). Before the instructions come from the central nervous system to move a muscle, nerves sur-rounding that muscle store packages of ace-tylcholine in tiny spherical compartments or vesicles. When the signal is ready to be sent to the muscle, these vesicles break through the membrane and burst, releasing a lot of acetyl-choline all at once. The neurotransmitter is then received by specialized proteins on the surface

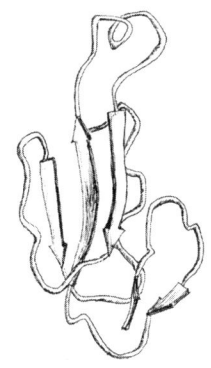

The structure of the cobratoxin protein (PDB code 1UG4) shows the characteristic "three-finger" arrangement, with a longer "finger" in the middle and two shorter ones at each side.

of the muscle cell known as cholinergic receptors. These receptors de-tect the presence of acetylcholine in the small gap between the nerve and muscle cells and transmit the signal to the muscle cell, activating cellular responses at lightning speed.

One of the major responses to the reception of acetylcholine by muscle cells is the release of ions, especially positively charged calcium, which triggers a coordinated muscle contraction. Without the release of acetylcholine from the nerve or its reception by cholinergic recep-tors on the muscle cells, the signal would not go through, and a muscle would not contract, no matter how much the central nervous system tried. What makes cobratoxin deadly is its ability to bind to cholinergic receptors mainly on the muscle cells, preventing acetylcholine from transmitting its signal. This causes paralysis in victims, beginning with drooping eyelids and a loss of the ability to move facial muscles. Then, as the venom makes its way to the muscles that control the lungs, breathing becomes difficult and death from respiratory failure can occur.

Cobratoxin is only one of many snake venom proteins known collectively as alpha neurotoxins.[7] All alpha neurotoxins affect the handshake between acetylcholine and its receptor, blocking the signal between nerves and muscles and ultimately leading to paralysis. Each of the toxins in snake venom has its own way of affecting its victim. The deadly black mamba, native to parts of central Africa, releases a unique combination of proteins, including its own version of alpha neurotoxin. Like cobratoxin, the black mamba's alpha toxin blocks acetylcholine signals, causing paralysis. But along with the alpha toxin, the black mamba also secretes other proteins known as dendrotoxins, members of the Kunitz-type protease inhibitor family.[8] While alpha neurotoxin is busy paralyzing its black mamba's prey, the dendrotoxins inhibit the prey's enzymes, which are trying to fight back by breaking down the alpha neurotoxin. The dendrotoxins in effect act as bodyguards for the alpha neurotoxin, blocking the prey's defense system. But they also have their own powerful effect on the nervous system. Since nerves transmit signals by allowing positive or negative charges to move through ion channels, anything interfering with the function of ion channels affects the way that nerves function. The dendrotoxins produced by the black mamba bind to specific ion channels, preventing the flow of potassium. The combined actions of the alpha neurotoxin and the dendrotoxins make the black mamba one of Africa's deadliest snakes.

In the tropical forests of India, the banded krait scours the forest floors at night looking for prey. The snake displays a distinctive alternating pattern of yellow and black bands across its body, which can reach up to seven feet in length. With a deadly cocktail of toxins, the krait packs a poisonous punch, which it uses to feed on other snakes.

The mixture of toxins produced by the krait venom contains an alpha toxin as well as a highly potent poison known as beta-bungarotoxin. Named after the Latin word for the snake, *bungarus,* this protein is an enzyme with a destructive nature. Beta-bungarotoxin belongs to a large class of beta neurotoxins, enzymes known as phospholipases.[9] These enzymes are extremely effective at breaking down the cell membrane.

Phospholipase enzymes are ubiquitous throughout life. We even have our own phospholipases, which we use to reorganize and recycle our cell membranes. Their action is usually highly regulated, only turned on when a specific repair needs to be made. But when the krait injects its venom into a prey, the phospholipase activity of the beta neurotoxin is unleashed without control. Immediately, the membranes of many cells begin to break down, disrupting their chemical composition and impairing their ability to function properly. As if this was not enough, beta neurotoxins cause nerves to release their entire store of acetylcholine all at once. This drives a massive response by the muscles. Overwhelmed by the flood of acetylcholine, cholinergic receptors become desensitized and unable to respond properly to the signal, resulting in paralysis.

The paralyzing effect of beta neurotoxins is much more potent and harder to reverse than that of alpha toxins. This is because the breakdown

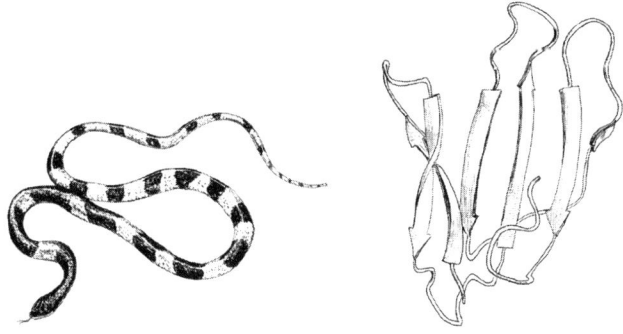

The banded krait produces a cocktail of deadly proteins including a powerful toxin belonging to the family of three-finger neurotoxins (PDB code 1F94), which disrupts the signal between nerves and muscles of a prey.

of the lipid membrane blocks the ability of the nerves to make more acetylcholine. Unable to replenish the acetylcholine supply, the nerves are helpless when the brain sends the command to contract a muscle. The communication system between the nerves and the muscles completely, and often irreversibly, shuts down. Lungs stop inflating, and the heart stops pumping, as the prey is devoured by the snake.

Perhaps the most dangerous snake in the world lives in the deserts of the Australian outback. The inland taipan packs a concentrated concoction of some of the deadliest proteins on the planet, with one bite delivering enough venom to kill a hundred people.[10] In addition to having its own alpha and beta neurotoxins, it possesses proteins that each act on a different part of the body in the most sinister of ways.[11] Some of these injected proteins pre-digest the victim from the inside out. An enzyme toxin known as metalloprotease begins to chew up the muscle tissue of the prey, while another, known as hyaluronidase, breaks down the surrounding connective tissue, accelerating the spread of the venom. If that wasn't horrifying enough, another class of toxins comes in to finish the job. Hemotoxins affect the blood of the victim, which is usually what leads to its death. When exposed to small amounts of blood, hemotoxins cause rapid clotting that turns the blood into goo. But within the larger circulatory system of an animal, these proteins actually have the opposite effect. As thousands of tiny blood clots form around the hemotoxins, the proteins required for clotting become used up, leaving the rest of the blood without any defense against bleeding. The victim then loses excessive amounts of blood from the site of the snake bite.

Hemotoxins are some of the most powerful weapons carried by snakes. These deadly proteins make it easier for a predator that lacks arms, legs, claws, or talons to subdue its prey. Many vipers, in particular, rely heavily on powerful hemotoxins to slow down their prey and to make them easier to devour. Neurotoxins and hemotoxins work together to paralyze the prey and put stress on its circulatory system, which causes a sharp drop in blood pressure as the animal collapses. Meanwhile, tissue surrounding the bite begins to die in a process known as necrosis. Severe hemorrhaging begins, accompanied by excruciating pain and ultimately death.

How snakes came to possess such powerful venom proteins is still not quite clear. Early snakes grew to massive sizes and developed strong muscles to capture prey and strangle them to death, much like today's boa constrictors. But as the climate changed, snakes, like most other reptiles on the planet, grew smaller and expanded into more arid environments, which required a new set of hunting skills for survival. Snake venom may have developed as early as 200 million years ago when a set of genes that had a different function in the reptile became duplicated, providing other copies that could undergo mutations without affecting the original genes.[12] The new copy created by this process of gene duplication would have provided a huge survival advantage to its carrier. As the gene coding for a protein toxin was passed down from generation to generation, new functions appeared, giving rise to the wide variety of snake toxins we see today. But just as snakes were gaining this survival advantage, a group of mammals, in-

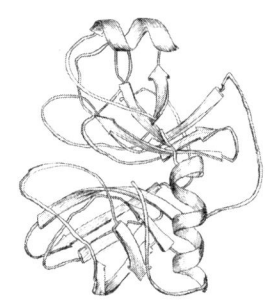

The venom of a viper packs a deadly mix of potent proteins, one of which is an enzyme known as serine protease. This protein from Russell's viper (PDB code 3SBK) is used to predigest its prey.

cluding our own ancestors, were going through their own evolution. Following the mass extinction of large dinosaurs around 65 million years ago, small mammals eventually gave rise to primates with larger brains, forward-facing eyes, and new ways of communication. The very proteins found in snake venom may have directly contributed to our own evolution.

Studies of both humans and other primates have shown that there is a built-in alarm system that alerts us to the presence of snakes.[13] When shown blurry pictures of different types of dangerous animals, both humans and monkeys are able to perceive the image of a snake more easily than any other type of predator. The image of a snake stimulates a region of the brain called the pulvinar, a highly developed center in primates used to detect relevant images within a context. The same re-action to blurry snake images is seen in children and even in primates that have never seen a snake before. Anthropologist Lynne Isbell has proposed the "snake detection hypothesis," which suggests that pri-mates have an intrinsic awareness of the deadliness of snakes, a kind of inherited memory of our ancestors' lethal encounters with these venomous reptiles over millennia, since our survival may have been dependent on an innate ability to quickly recognize anything that re-sembled a snake. Today it is estimated that around one-third of all humans live with ophidiophobia, an extreme fear of snakes. Our col-lective fear of snakes originated from the toxic nature of the proteins in their venom, which likely steered our evolution. It is no wonder that most of us will startle when our eyes catch a glimpse of a twisted twig or a garden hose resembling an ancient killer.

Snakes have truly perfected the use of proteins as potent compo-nents of a deadly venom, but they are not the only reptiles that produce toxins. In the tropical islands of Indonesia, the largest and perhaps scar-iest of all living reptiles roams the beaches. Komodo dragons are fierce predators that resemble ancient dinosaurs. For a long time, Komodo dragons were thought to kill their victims by introducing deadly bac-teria trapped in their teeth and released when they bite. The bacteria would then infect the flesh of an animal, killing it within a few days. But several toxin proteins have been identified in the giant lizard's sa-liva, some of which work in a similar manner as the hemotoxins used by vipers.[14] These proteins prevent blood clotting at the site of the wound and send the prey into shock, eventually shutting down multiple organ systems in its body. This is accompanied by a sharp drop in blood pressure, and as the prey is immobilized, the Komodo dragon moves in to devour it.

Not all deadly creatures have left such a scary image in our collec-tive mindset; some of the deadliest protein toxins in the world are not even produced by animals. Plants have also evolved ways to fend off predators by building an arsenal of toxin proteins.[15] One of the most notorious examples is a deadly protein, known as ricin, that is found in castor oil plants.[16] Ricin's toxicity lies in its ability to block the machinery responsible for making proteins. When consumed by an unsuspecting victim, ricin finds its way into the cells of the host by binding to the cell surfaces and hijacking the transport pathways used to bring cargo inside the cell.[17] Once inside, ricin latches on to the ribosomes, the large machines responsible for translating genes into functional proteins. Without the ability to make proteins, the entire cell shuts down, unable to sustain itself or proliferate. Ricin is extremely toxic; as little as one gram of it can kill an adult human.

Ricin belongs to a family of plant protein toxins known as ribosome inhibiting proteins (RIPs). More than a hundred different plant species make RIPs to defend against insects and pathogens such as viruses and other parasites. In addition to RIPs, many plants produce proteins known as lectins. Defined by their ability to bind tightly to sugars, lectins can have a variety of effects when consumed. Lectins found in common legumes, like beans and lentils, are harmful to us if consumed raw, but the heat from cooking denatures them and renders them harmless. Lectins from the castor oil plant, by contrast, can be quite dangerous. Even with heat treatment, they can latch on to the sugars decorating the surface of multiple red blood cells and link the cells together in a clump resembling a blood clot. This can have profoundly harmful effects if the lectin protein finds its way into the bloodstream. Clumps of red blood cells block tiny arteries and prevent blood flow to vital organs.

BEAUTIFUL BUT LETHAL

The underwater paradise created by the shallow reefs of the Atlantic and Pacific Oceans is home to some of the world's most beautiful, and colorful, organisms. Elaborate fish show off yellows and blues, while corals glow pink and green. Along the sandy bottom of the shallow reefs, a living relic slowly moves. A sea snail hides its soft, vulnerable body inside an elaborately painted spiral shell constructed within the confines of the perfect golden ratio. Patches of brown spatter the creamy white background on the surface of the snail's nearly indestructible exterior. Only in the 1950s were cone snails discovered to be venomous. One day, Alan Kohn, an undergraduate student at Yale University, placed a cone snail in a tank with a small fish. What he saw came as a shock. The curious fish approached the cone snail's tail-like protrusion, then in a flash, the cone snail stunned the fish and began swallowing it whole.[18]

The tail-like protrusion Kohn observed is known as the proboscis. Inside this long tube is a sharp tooth that acts like both a harpoon and a hypodermic needle. As the proboscis latches on to the prey using suction, its hollow tooth is injected into the victim, allowing the snail to administer a cocktail of protein toxins while barbs on the tooth keep it latched onto the doomed animal. Within less than a second, the prey begins to feel the effects of the array of toxins, first briefly convulsing and then becoming paralyzed. That's when the snail reels in its meal and swallows it completely. The harpoon-shaped tooth, also used by the snail to snare fish, worms, or even other snails, is quickly discarded after the meal is digested. Then a new harpoon is loaded from a quiver stored inside the snail.

There are nearly a thousand species of cone snails scattered across the warm waters of coral reefs and shallow seas near the tropics. Using a similar hook-and-line technique to hunt, each species has concocted its own deadly mix of toxic proteins, with each toxin adapted to different types of prey.

Two decades after Kohn made his discovery, two undergraduate students, Craig Clark and Michael McIntosh, began working in the lab of Professor Baldomero Olivera at the University of Utah, unaware that they were about to make a groundbreaking discovery.[19] Professor Olivera had given them one of his side projects; they were tasked with extracting and analyzing the venom of the *Conus magus*, a cone snail whose name means the "magician cone." By then, it was already known that the cone venom contained at least a handful of small proteins that could kill a mouse when injected into its bloodstream in the lab. Clark and McIntosh began by separating the different components of the

guard is down, for example when we have a burn or a wound, it can cause serious infections that are difficult to treat. Instead of living in harmony with our bodies, the bacteria turn into enemies. When this happens, *S. aureus* has many protein weapons at its disposal. One of these, alpha toxin, has quite the stunning structure.[28] It was the first identified member of the "pore-forming beta barrel" family of toxins. *S. aureus* alpha toxin consists of seven protein chains. Together, they form a large hollow barrel structure that creates a pore or a hole in the membrane of a human cell, mostly red blood cells. These pores, called beta barrel pores, cause cell death as the insides of cells leak out through them.

Many species of bacteria produce a wide variety of toxins, which can make us sick or even kill us. Another natural skin dweller is a species of bacteria known as *Pseudomonas aeruginosa*. Most of the time, it is completely harmless to us. But some strains can harbor deadly proteins that they unleash when given a chance. *P. aeruginosa* infections are extremely common in burn patients who have lost significant amounts of skin, the first line of defense against most bacteria. Pathogenic strains are also the main culprits of infections in immunocompromised individuals, or patients with cystic fibrosis. Like *S. aureus*, they are also very commonly present in hospital infections, which can be notoriously difficult to treat. *Pseudomonas* can produce a protein toxin called exotoxin A, which disables our cellular machinery responsible for making proteins.[29] The exotoxin made by *P. aeruginosa* puts immense stress on the demand for functional proteins and can kill entire cells.

Explaining why bacteria go from getting along with humans to turning into a major infection is not straightforward. But the human body is an extreme environment. Our skin and the bacteria on it are exposed to the elements. Inside our gut, bacteria are exposed to a myriad of chemicals, digestive enzymes, and changes in pH. Even our immune system can become dangerous to bacteria that our body normally recognizes as friends. Any shift in our bodies can trigger peaceful bacteria to suddenly wage war against their home. For instance, a burn or wound on the skin activates our immune response, which in turn triggers bacteria such as *S. aureus* to use their normally silenced toxins to fight against us. People with compromised immune systems are all too familiar with these battles. Their cells are constantly trying to survive the onslaught from bacteria as friendly neighbors turn into deadly pathogens.

Beyond the natural skin-dwellers, some bacteria are far more sinister, producing some of the most toxic proteins known to science. Perhaps the deadliest toxin on Earth is produced by common soil-dwelling bacteria. *Clostridium botulinum* is a type of anaerobic bacteria that grows just below the surface of the soil. This oxygen-free environment allows it to grow and thrive, and when food is scarce, the bacterium has the astonishing ability to change its entire metabolism to produce a strong shell within which it can lay dormant indefinitely.[30] *C. botulinum* can be found in the soil of backyard gardens, and sometimes it finds its way into canned foods. If not heated properly to completely kill dormant bacteria, the oxygen-free environment inside a tightly closed Mason jar provides the perfect condition for the bacterium to come out of its shell, grow, and produce its toxin. Each year roughly two hundred people in the United States become infected with botulism from consuming improperly canned foods or unpasteurized milk or honey. The botulinum toxin produced by the bacteria, like the neurotoxins from snake venom, interferes with the signal between muscles and nerves. The toxin blocks the release of acetylcholine from the nerves, the signal that tells the muscles to contract. As a result, paralysis occurs, and patients die from respiratory failure as they become unable to expand their lungs. This toxin is considered the most potent and deadliest substance known to date. Even as little as one ten-millionth of a gram can kill an adult human if it finds its way into the bloodstream.[31] But like other deadly proteins, it now has various medical applications. As a powerful muscle relaxant, the toxin is used to treat severe muscle spasms as well as conditions such as cerebral palsy.[32] It is also used to manage symptoms of recurring migraines.[33] Perhaps most famously, the botulinum toxin is used in the cosmetics industry under the brand name Botox to smooth out wrinkles and erase other marks of aging by relaxing facial muscles.[34]

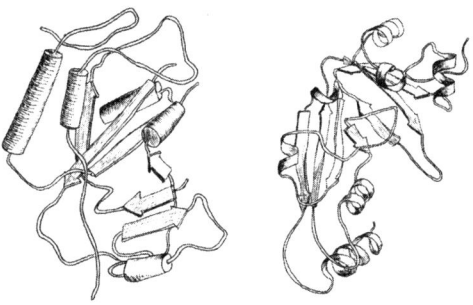

A segment of the diphtheria toxin (*left;* PDB code 1DDT) and a segment of the exotoxin A from *Pseudomonas* (*right;* PDB code 1IKQ); both are bacterial toxins. They act as powerful blockers of the cellular machinery needed to make proteins in humans.

The botulinum toxin is unique in many ways. As a neurotoxin, it works only on higher organisms with advanced nervous systems. It has no effect on other species of bacteria, and so it can't help its own botulinum bacteria compete with other bacteria for resources. But it is amazingly stable. The toxin can remain unaffected by the environment or the immune system long after the bacteria carrying it are dead, and even as it passes through the highly acidic environments of the host's digestive tract. (This means that poisoning from consuming contaminated canned goods can happen months or even years after the bacteria that produced the toxin die.) Moreover, many precise steps must take place in a specific order for the toxin to be activated after its release from the bacteria. First, the toxin is absorbed by the digestive tract and distributed to the different parts of the body through the circulatory system while evading the host's immune response. The toxin then finds its way into nerve endings by binding to a type of lipid, most commonly found on the outside of nerve cells. This is when a clever Trojan horse mechanism becomes activated. A segment of the toxin binds to a receptor on the surface of the nerve cells, which mistakes it for an essential nutrient and brings it inside the cell. Once inside, the toxin targets a set of proteins known as SNAREs for destruction. SNAREs are essential for mediating the release of the neurotransmitter acetylcholine, the signal released by nerves to activate muscle contraction. Without properly functioning SNAREs, the release of acetylcholine becomes impaired, and with it, muscle contraction.

It is still not clear why *C. botulinum* produces such a powerful toxin or why such a complex mechanism for toxicity has evolved.[35] Most toxins made by bacteria provide some survival advantage, either helping them to colonize their host, suppress the body's immune response, or compete with other bacteria for resources. One possible explanation is that having such a potent toxin could help spread the bacteria through the bodies of dead victims, but there is little evidence for that, mainly because the toxin can outlive its producer. An alternative explanation is that the toxin may play a different role in bacterial metabolism, one that happens to make it a potent human toxin. But this too has little supporting evidence since many strains of *C. botulinum* survive and thrive without producing the toxin. Why a soil-dwelling bacterium needs to make this powerful toxin remains a mystery, one that highlights the intricate and sometimes unexpected ways in which evolution shapes the natural world.

WHAT IS A TOXIN?

Just as the fight for survival rages on in the arid Sahara and the desolate expansive desert of our skin, similar biological warfare has transpired in the waters of the Nile. The river is home to countless fish and aquatic organisms that have lived there for millions of years. The Nile tilapia is one of the iconic residents of the mighty river. It has helped shape the culture of peoples near its shore and sustained countless generations of people living along this longest river in the world. But the tilapia is not the only resident of these fresh waters. Microscopic organisms thrive in the Nile's rich water, often producing their own protein toxins. Without a defense system, the tilapia would succumb to infections by bacterial invaders that accompany the plankton and algae they feed on. Like humans, tilapia benefit greatly from their own gut bacteria—the "good" bacteria—which help them digest their food and replenish their supply of essential vitamins. Beyond aiding in all organisms' gut health, these bacteria also provide support and protection in other parts of the body. For example, human skin has its own unique microbiome that possesses remarkable abilities to neutralize or metabolize harmful substances that we encounter. Even many plants require specific bacteria in the ground to help break down organic material that promotes growth as well as to fight off predators or other competing plants. But among all these good bacteria, danger always lurks. Other types of bacteria can produce toxins if allowed to grow and thrive.

Tilapia are armed with a set of weapons to help defend against infections. One of the most effective weapons the fish have against bacterial invaders is a set of proteins known as antimicrobial peptides, or AMPs. These small proteins are produced in the gut of the fish, and when there is an infection, they are secreted to where they can target the bacteria for destruction.[36] AMPs are known as piscidins, meaning "fish toxins." These proteins have a high affinity for the membrane of bacteria. They quickly latch onto the surface of bacteria and poke holes into their membranes, causing the insides to spill out and ultimately killing the tiny invaders. To us, they are perfectly harmless. Many popular dishes in Egypt and all around the world feature tilapia, prepared in a variety of ways, all safe to eat and each more delicious than the next.

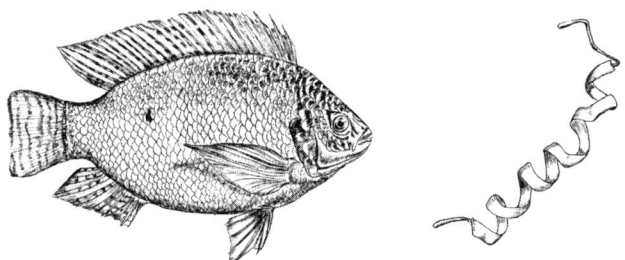

Many animals produce proteins known as antimicrobial peptides (AMPs) to fight against bacteria. Tilapia produce a type of AMP known as piscidin 4 (PDB code 5H2S), a small protein that can poke holes in bacterial membranes.

Piscidins are not the only example of powerful bacteria-killing toxins. In fact, AMPs are a large superfamily of proteins produced by a variety of organisms.[37] They have one main goal: to kill bacteria. Even invertebrates like fruit flies and silkworms produce a type of bacteria-killing AMP, known as cecropins. Much like fish AMPs, cecropins target the membrane of bacteria for destruction, allowing the insects to escape deadly infections in a world teeming with bacteria. One insect in particular makes its living in some of the most bacteria-infested places in the world. The flesh fly feeds on corpses, dead tissue, and the infected wounds of animals. Living in such close proximity to unimaginably high numbers of bacteria, the flesh fly produces powerful AMPs to protect itself from infections. The aptly named sarcotoxins, meaning "flesh poisons," are a class of AMPs produced by the flesh fly. Sarcotoxins are highly effective at killing bacteria by breaking up their membranes. The evolution of AMPs and their prevalence in so many living organisms has made surviving in a world filled with countless bacteria a possibility. These toxic substances have, in effect, created a balance in the biological arms race among different species, allowing each one to carve out a unique niche.

With their natural ability to kill bacteria, many AMPs are now being investigated as much-needed new antibiotics, especially with the rapid rise worldwide of antibiotic resistance.[38] In fact, several AMPs have already been clinically approved for treatment of infections when traditional antibiotics have failed. As promising as AMPs are in the fight against antibiotic resistance, many challenges lie ahead. Because AMPs are proteins, if taken by mouth, most of them will be digested by our gut enzymes, rendering them inactive. Moreover, proteins do not typically move easily from the digestive tract into the bloodstream, where they could fight a bacterial blood infection. At the same time, there are potential risks to injecting AMPs directly into the bloodstream of a patient. AMPs could be recognized as foreign invaders by the patient's immune system, potentially leading to a massive and even life-threatening immune response known as anaphylactic shock. The use of AMPs as topical treatments applied directly to infected wounds or burns may be a promising move forward, however. AMPs applied on skin burns have already helped prevent infections and improve recovery. Recent studies in mice have also shown that applying AMPs to catheters greatly reduces the rate of urinary tract infections. As more and more AMPs are discovered across all the kingdoms of life, the potential for their use against infection has become more of a reality, offering renewed hope in the battle against the global rise of antibiotic resistance.

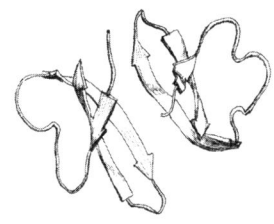

Almost all organisms on the planet produce some mixture of proteins to fight against the immense number of bacteria in the environment. We, humans, are no exception. We produce a cocktail of toxic antimicrobial peptides known as defensins.[39] As the name suggests, defensins act as our first line of defense against bacteria. And naturally, defensins are produced in places where bacteria are most likely to come into contact with our bodies. Our skin cells produce massive amounts of defensins to limit the growth and proliferation of bacteria. Similarly, the mucous membrane of the tissues in our lungs produces defensins to protect us from the bacteria we take in every

Our innate immune system is hardwired to fight infections. One of the weapons we use to fight bacteria is an antimicrobial peptide (AMP) known as defensin alpha 1 (PDB code 2PM4). This small protein, which helps defend against bacteria, is found along our digestive tract as well as on the surface of our respiratory tract.

time we breathe. Our defensins are part of what's known as the innate immune response, a collection of proteins pre-programmed into our genetic code to indiscriminately kill invaders. They are distinct from other immune proteins like antibodies, which are components of the adaptive immune response. Whereas antibodies are adapted to each invader's specific shape, defensins are blanket killers. They are potent poisons against a wide range of bacterial species. And like most antimicrobial peptides, defensins act by disrupting the membrane of bacteria, causing them to burst. If we relied solely on antibodies to protect us, our immune system would be working overtime to kill all the different bacteria we encounter every day. Having defensins act as the "catch-all" of our immune response to bacteria ensures that our bodies are always prepared. Without these potent protein toxins, life in this world filled with hostile bacteria would be nearly impossible.

It is easy for us to call snakes and scorpions venomous because they can kill us with their powerful toxins. But at the protein level, we are all killers. Each day, countless bacteria fall victim to our murderous proteins, such as the toxins we produce on our skin, in our saliva, and on the surface of our respiratory tissue. In a world where survival requires killing, all organisms can be considered toxic on some level. In the theater of life, murder is the plot and proteins play the lead role. Ultimately, as the curtain closes, death comes to us all, often at the hand of a protein.

9

DYING

MAGGIE M. FINK

I KNEW SHE WAS going to die—amyotrophic lateral sclerosis (ALS) is a death sentence—but Mary Inez didn't seem to mind. She didn't know when she was diagnosed in 2004 that she would have only nine months left with us. I don't remember much of those months after her diagnosis. Everything seemed the same except we made the two-hour drive to Anderson, Indiana, every weekend instead of every other month. The same French toast sticks from the Schwann man were still there, the same mugs of Folgers instant coffee, the same bedspread I had slept on since I was a baby, the same photos of my mom and aunt covering all the walls. There was something depressing in the air, but we all still laughed and ate and got yelled at by my grandfather for "wrastlin' too loudly." Death was knocking on the door of my grandmother's mid-century-modern, blue-carpeted, wood-paneled house, and we all ignored it for as long as we could.

Even when we knew we had to at least let death come and have a seat at the table, the absurdity of its presence made everything a bit funnier. Grandma would sneak us twenty-dollar bills so Grandpa wouldn't know, even though it was obvious that a starving, eighty-pound woman who couldn't walk on her own was taking money from his wallet. She demanded my mother drive us all to her favorite bank forty-five minutes away to purchase stamps, just a week before she died. And even though she couldn't eat, she wanted Wendy's chili brought to her, which would stay uneaten on her TV tray by the chair she now lived in.

I wasn't there when my grandmother died. But my mom told me that even on that day, the humor of death was still there. Grandma was no longer awake. She was resting in her chair, quietly breathing. But every so often, the footrest on her corduroy 1970s lounge chair would shoot up, flinging her body violently as her bereaved family was gathered around her. A delightful terror. She had called me the night before she died. I couldn't understand anything she was trying to say except "I love you," and even that I might have made up in my head. Those nine months still seem like an eternity to me. The nine months I carried my own children seemed like an eternity. And the end of both seems blurry and fabricated in my mind to suit some narrative of the sacredness of it all. But really, there is nothing sacred about a grandmother dying and a child being born. This is life.

Death and life. Life and death. This is the tempo that dictates our lives. Sometimes death comes a beat too soon, or it holds on to the final note too long, refusing to exhale. But the ending is always the same. This surety means that every part of our bodies, each cell, goes through this couplet of life and death. Cells have developed ways to avoid death from toxins, infections, and other threats outside that could harm us, but many of the most devastating diseases and disorders that hasten the final notes of death come from the processes in our own bodies going awry. If you consider the innumerable events that must occur at the molecular level every second of every day just to keep us alive, it is astounding that we are alive at all.

We can thank proteins for carrying out the chaotic processes of life. These tireless machines are nearly perfect at executing their jobs, and even if mistakes are made, typically they can be ignored because other proteins pick up the slack without missing a beat. Some of these mistakes occur simply because a protein doesn't adopt the correct three-dimensional structure and becomes a misfolded protein. This isn't always inherently dangerous to the cell. Usually, misfolded proteins are quickly escorted away and broken down so that they can't become toxic to the cell. But sometimes that is not enough. Diseases like Alzheimer's, Huntington's, and ALS are all caused by proteins that don't fold into their proper 3D shape and are unable to be cleared out of the cell, which causes other proteins around them to unfold. As more and more of these misfolded proteins start to aggregate in the cell, they cling together, with devastating consequences for the cell's structure and function.

In Alzheimer's disease, two proteins are closely associated with the damage in the brain that causes the characteristic cognitive decline: amyloid beta (AB) and tau.[1] Amyloid beta is just a small portion of a larger protein known as amyloid-beta precursor protein, which plays a pivotal role in orchestrating cellular activities crucial for our cognitive well-being.[2] Imagine this protein as a molecular architect, meticulously crafting the scaffolding that maintains the structural integrity of our neurons. Under normal circumstances, this protein is cleaved from its precursor, a larger initial structure with amino acids attached to one end that keep it from functioning properly. In a precisely choreographed manner, the protein is chopped up to form shorter fragments that contribute to the formation of synapses—the bustling junctions where nerve cells exchange vital signals. In essence, amyloid beta is a foundational player in synaptic function; it maintains the harmonious rhythm of our cognitive orchestra.

While its cellular function is not fully understood, the role of amyloid beta in Alzheimer's is clear. Although they are typically folded with precision, AB proteins can deviate from their intended structure, especially as a person ages or experiences significant stressors. These misfolded proteins become the instigators of a biochemical upheaval and set the stage for the cruel symptoms of the disease.[3] Missteps in this folding process result in structural deformities. Specifically, AB begins to unravel its helical structure and misfold into a construct rich in beta-strands, which form a molecular quilt. Individual strands become woven together in a precise and orderly fashion, creating a flat and expansive sheet. In properly folded proteins, beta-sheet arrangements usually impart strength and stability by forming a robust backbone reminiscent of a well-organized lattice; when ABs change into beta sheets, however, they become "sticky" and cause other normally folded ABs to adopt the rogue beta-sheet structure. That's when the cascade begins. The misshapen proteins aggregate, forming amyloid plaques, the ominous hallmarks of Alzheimer's pathology. Far from the architects they once were, these misfolded AB proteins become tangled debris that interfere with the delicate signals between neurons, leading to the familiar symptoms of dementia and memory loss.

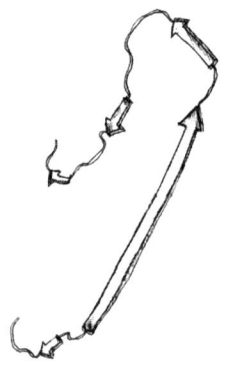

Tau misfolding is a key event in the development of neurodegenerative diseases such as Alzheimer's. Normally, the tau protein has several short helices and long regions with no secondary structures. The misfolded tau, depicted here, has abnormal beta-sheet-rich structures that lead to the formation of tangles (PDB codes 2MZ7, 8WCP).

The other protein commonly found in tangled clumps in the brains of Alzheimer's patients is tau. Normally, tau helps form long-term memories, facilitates communication along axons, and promotes microtubule assembly and stability. While tau can be found in other places throughout the body, neurons and other cells along the central nervous system have an abundance of it. This localization of the tau protein means that when it starts to go rogue, cognitive function can be dramatically altered. Unlike ABs, however, which unfold and cause plaques to form, tau experiences significant chemical modifications that cause it to become sticky. Under normal conditions, tau can undergo a process called phosphorylation, in which amino acids on the surface of the protein are decorated with phosphorus and oxygens, disrupting its role in microtubule organization. (This is yet another example of how proteins can modulate their shape to change their function.) The phosphorylation signals to tau and other proteins to move around, bind to something new, or halt altogether. In Alzheimer's, tau becomes hyperphosphorylated. Too many phosphates added to the surface of the protein cause it to let go of the microtubules. These free, hyperphosphorylated tau proteins begin to aggregate, much like how amyloid proteins clump into misfolded knots. Not only do these tau tangles damage cell tissue, but the loss of functioning tau needed to regulate memory and cognitive function also contributes to Alzheimer's symptoms.[4]

Many families must watch a loved one with Alzheimer's decline over many years. In fact, Alzheimer's is the most common cause of dementia in the United States. Other diseases caused by protein misfolding are less common, but no less devastating. ALS, which killed my grandmother, is one of these; instead of triggering memory loss, however, ALS causes a decline in muscle function, leading to paralysis, an inability to eat or speak, and ultimately the loss of vital bodily functions. At the heart of this neurodegenerative disorder lies a tangled web of proteins gone awry, much like what is observed in Alzheimer's.

ALS is intricately linked to the misfolding of proteins, notably a protein called superoxide dismutase 1 (SOD1).[5] In a healthy cellular ensemble, SOD1 plays the role of the guardian. It protects nerve cells from damage caused by unstable oxygen molecules. But in ALS, this protein undergoes a transformation akin to a once-reliable ally turning traitor. The misfolded SOD1 adopts a toxic conformation, setting off a cascade of events that eventually lead to the demise of motor neurons, the nerve cells responsible for muscle control. It's as if a key instrument in the grand symphony of the body starts playing off-key and the entire orchestra goes into disarray. A single rogue SOD1 causes nearby, properly folded proteins to unfold, disrupting the chemistry that keeps them functioning, and turning them into toxic molecules.[6] A misfolding frenzy ensues, and clumps of proteins accumulate within cells. As these toxic protein aggregates spread through the nervous system, the once-harmonious communication between neurons breaks down. The biochemical unraveling of ALS is a tragic saga in which proteins, the molecules of life, inadvertently become the agents of cellular chaos, leading to the silent but profound erosion of a person's ability to move, speak, or breathe.

Both Alzheimer's and ALS are the result of proteins becoming defective in a way that affects their structure. Regardless of how these proteins begin to unravel, they all behave like an infection, spreading their deformity to other healthy proteins they come near. This behavior raises an interesting question: If a single protein can spread its diseased state to another protein, essentially replicating itself, could these neurodegenerative diseases be contagious? Could a protein, with no genetic material, act like a virus or a bacterium?

A NEW ERA OF INFECTION

In the remote highlands of Papua New Guinea in the 1950s, the air was thick with death. A disease had taken hold of the Fore people; men, women, and children trembled on the ground. Bursts of manic laughter occasionally escaped their lips. Then inescapable death; the disease had a 100 percent fatality rate. They called the disease negi-nagi, from the Fore word for foolish person, because of the creepy laughs from the dying. The tribe believed the disease was caused by ghosts or evil spirits. Soon government officials took note, and two researchers were sent to investigate what was causing this deadly laughing sickness. One of those men, Daniel Carleton Gajdusek, was a virologist. For years, he lived with the Fore, observed their customs, and learned their language to try to understand the root of this disease.[7] While other doctors and scientists were certain it was a genetic disorder specific to the Fore—the result of intermarriage—Gajdusek believed he had stumbled on an entirely new infectious agent.

Soon after he arrived, Gajdusek took note of something distinctive about the Fore: they practiced cannibalism. They would consume those who had died, digging up the bodies after a few days of being in the earth. For them, this was a sign of respect for the dead loved one, an important part of their culture, akin to communion. But Gajdusek suspected this ritual might be linked to the kuru, the name researchers had given the disease. It was clear that kuru was able to spread, but it was different from other infectious diseases. It didn't discriminate against the young or the old, so it wasn't like other neurodegenerative diseases that usually affect the elderly. It also lacked the typical inflammatory markers seen in transmittable diseases.[8]

The cannibalism ritual itself was possibly a clue. Though men, women, and children all suffered from kuru, women and children seemed to catch the disease much more often. In the cannibalism ritual, the women and children usually ate the brains of the deceased, while the men preferred the muscles. Gajdusek predicted that whatever was causing this devastating disease must be in the brain. After returning to the United States, Gajdusek began conducting experiments at the National Institutes of Health. Using tissue from the brains of people who had died from kuru, he infected chimpanzees and waited to see if they developed kuru. After two years, one did, proving that something in the brain of an infected person could cause the same disease in someone else. It wasn't a virus, but something completely new.

The long incubation time of kuru meant Gajdusek's research took many years. It wasn't until almost twenty years after he first visited the Fore tribe that other researchers were able to show that this new disease was not caused by a virus but rather by a protein.[9] Moreover, kuru might not be the only disease caused by proteins. Two researchers, Tikvah Alper and John Stanley Griffith, had each been studying a similar disease in sheep: scrapie. Though they worked at different institutions in England, they had observed similar characteristics to that of kuru—the long incubation period, the lack of an immune response, neurodegeneration, and always death. Their research showed that transmission of the disease occurred even when the tissue was treated with UV light, a process that destroys DNA. This indicated that whatever the infectious agent was, it didn't contain any genetic material. Combined with Gajdusek's work, it suggested that a protein was responsible for the devastating disease. Eventually, in the 1980s, the final piece of the puzzle came together when Stanley Prusiner at University of California, Berkley isolated the protein responsible for these diseases.[10] He called them prions, short for proteinaceous infectious particles. Simply adding this specific misfolded protein to cells was enough to cause the disease now known as transmissible spongiform encephalopathy. With these discoveries over several decades, the existence of contagious proteins was accepted, and both Prusiner and Gajdusek won Nobel Prizes for their groundbreaking work on prions.

What makes prions particularly insidious is their ability to act like molecular puppeteers and manipulate other proteins into adopting their abnormal conformation, just like in Alzheimer's and ALS. The misfolded prion triggers a cascade of misfolding in healthy proteins, transforming them into replicas of the malevolent prions.[11] This domino effect leads to the formation of protein aggregates that cause the spongelike appear-

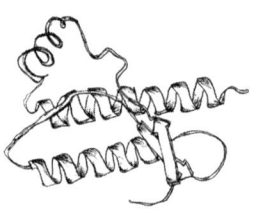

Prions are proteins that can trigger the misfolding of other proteins, creating aggregates that can lead to severe neurodegenerative diseases.

ance in affected brain tissue and, ultimately, the characteristic neurological symptoms observed in both animals and humans. Many people may have heard of mad cow disease, a form of this prion infection that was transmitted from beef contaminated with bovine spongiform encephalopathy. An outbreak of mad cow disease during the 1980s in Europe brought prion diseases to the forefront of international news. All of the 177 people who reportedly died from the disease had consumed contaminated beef products.[12]

Prion diseases don't always arise from consuming contaminated brains or cows. Properly folded prions occur in humans and animals, but their purpose is not well understood. As in other protein-folding diseases, a random misfolding event in a single prion can occur, triggering the unfolding of nearby prions. Scientists predict that this may have been what began the outbreak of kuru among the Fore people. A prion spontaneously misfolded, leading to a person's death. The body was then consumed by the tribe, and because the rogue protein was still active in the brain, it could spread its misfolding to other people, propagating the disease until it killed the next person, and so on. The original event most likely occurred at random, a spontaneous event that completely altered our understanding of the role of proteins in disease.[13]

Most prion diseases, as well as prion-like diseases such as Alzheimer's and ALS, occur randomly. No predictable signs. No screenings. Simply the body getting old and losing its ability to function properly, even at the cellular level. In a small number of instances, however, a genetic mutation is present that significantly increases an individual's chances of developing these diseases. These mutations change the DNA carrying the instructions for making the protein, which becomes more prone to misfolding. Because that genetic information is passed on to children, these diseases are hereditary. Though this happens in only a small number of cases for prion-like diseases, many other protein diseases are genetic, that is, the result of faulty instructions for making the protein being passed from one generation to the next.

Genetic diseases arise from mutations in the DNA sequence. When mutations occur, they introduce typos that disrupt protein synthesis. These errors may result in proteins with slightly skewed shapes or impaired functions. In other cases, more drastic mutations can cause a complete breakdown in protein production and leave cellular processes without their designated performers. The consequences range from subtle biochemical hiccups to catastrophic disruptions, depending on the nature and location of the genetic glitch.

Calling these diseases genetic is not technically inaccurate, but focusing only on the DNA diminishes one's understanding of the mechanism of the disease, which is ultimately the result of an impaired protein. Some genetic disorders arise from a mutation in a solitary gene that affects the structure or function of only one protein. Well-known single-gene genetic disorders include sickle cell anemia, cystic fibrosis, and Huntington's, but the list of known genetic disorders to date is between seven thousand and ten thousand.[14] Many of these are rare and fatal, even untreatable, while others are more common and exhibit a wide range of outcomes. It is remarkable that out of our approximately twenty-five thousand genes that code for proteins, a mistake in a single one can wreak havoc on the body. Investigating these single-gene disorders serves as a reminder of the intimate connection between our genetic makeup and the proteins that shape our physiological destiny. The entire biological drama of life can hinge on the performance of a single actor.

Cystic fibrosis is a common example of a single, seemingly insignificant protein disrupting a person's entire physiology. The disease is most common among white people, with one in 3,500 newborns being diagnosed with the disorder.[15] A mutation in the cystic fibrosis transmembrane conductance regulator (CFTR) protein causes the characteristic persistent coughing, difficulty breathing, and recurrent lung infections. But this protein's normal job is quite mundane. It sits embedded in the cell membrane as a channel, forming a tunnel from inside the cell to the outside world. Through this channel, CFTR regulates what comes in and out. Specifically, it moves chloride, a key component of salt, to the cell surface.[16] CFTR is primarily located in the lungs and important digestive organs like the liver and pancreas. All of these cellular locations rely on slippery mucus—a thick but limber fluid critical for moving digestive enzymes and for trapping and clearing out gunk that clings to these organs. When CFTR is functioning normally, bringing salt to the surface of the cell attracts water from the environment. This ensures that the mucus is the right consistency, that it's viscous enough to stick together but fluid enough to move easily. In the lungs, for instance, this process allows cilia—tiny hairlike structures—to clear away unwanted bacteria or viruses that could cause infections, and ensures optimal respiratory function. CFTR is a simple protein in the complex world of cells that, in healthy individuals, is taken for granted. Who would have guessed that a protein important for making mucus could be so critical for survival?

When a mutation occurs in the DNA coding for CFRT, chloride is unable to be transported to the cell surface. Salt, composed of sodium and chloride, cannot form; water does not accumulate; and a thick, viscous mucus is produced. The spectrum of CFTR mutations is vast, with over two thousand identified mistakes in the DNA.[17] These mutations range from a single amino acid being changed to deletions of entire sections of the protein and even extra amino acids. Each one can result in different functional consequences for the CFTR protein. Some lead to the production of a defective protein that fails to fold correctly and prevents it from reaching the cell surface where it's needed. Other mutations may allow the CFTR protein to reach the cell surface but impair its ability to open and close the chloride channel efficiently. Cystic fibrosis and its many mechanisms for disease highlight the intricacy of protein function and its dependance on accurate 3D structures to carry out day-to-day tasks inside our bodies.

While cystic fibrosis can be caused by a wide range of possible disruptions to the CFTR protein, other genetic diseases can be blamed on a much narrower set of mutations. Huntington's disease is one such genetic disorder that affects the brain, leading to progressive and irreversible neurological decline. At the core of this debilitating condition is a mutated gene known as HTT, which encodes a protein called the Huntingtin protein. This protein plays a crucial role in the normal function of brain cells, and alterations to its structure lead to the onset of Huntington's disease. It is involved in various cellular processes, including transport within neurons and cell-to-cell communication. Individuals with Huntington's disease have an abnormal expansion of a specific section of the HTT gene. The resulting mutated Huntingtin protein has an elongated section called a polyglutamine tract, which consists of repeats of the amino acid glutamine. This faulty protein causes a cascade of detrimental effects within the brain.[18]

In Huntington's, not every polyglutamine expansion leads to symptoms. At birth, most people have fewer than thirty-six of these repeats, which are necessary for proper protein function. As mutations cause that number to rise, however, the risk of developing Huntington's increases. There is a threshold where symptoms may or may not appear, and if they do, it will likely be late in life, with little time for an individual to develop the more progressive and severe symptoms.[19] But for someone whose Huntingtin protein contains more than forty of these repeats, it is nearly a guarantee that they will suffer from a more severe form of the disease. Even more devastating consequences are in store for individuals who have sixty or more repeats, nearly twice the number of a normal functioning protein. In these cases, the disease shows up very early in life, most often before a person turns twenty and sometimes as early as age two.[20] This early occurrence, called juvenile Huntington's disease, is cruel. After experiencing seemingly innocuous mood changes, young victims typically endure ten to fifteen years of cognitive and physical deterioration and suffering before succumbing to the disease.

As is true for many of these protein-based neurodegenerative diseases, the mutated Huntingtin protein tends to form clumps or aggregates.[21] These aggregates can accumulate inside nerve cells, disrupting their normal function and leading to cell death. The progressive loss of neurons, particularly in regions of the brain responsible for movement and coordination, contributes to the characteristic motor symptoms of Huntington's disease, which include involuntary jerking and difficulty with speech and balance. But the mutated Huntingtin protein doesn't affect only motor function; it also has widespread consequences for the brain's cognitive and emotional processing centers. The accumulation of toxic aggregates spreads to other regions of the brain, leading to disturbances in thinking, reasoning, and mood regulation. Individuals with Huntington's often experience psychological symptoms, including depression, memory loss, difficulties in decision-making, and difficulty regulating emotions, all of which contribute to the complexity and severity of the disease.

CFTR and Huntingtin are only two of the more than twenty-five thousand proteins in our bodies. As excellent as our molecular mechanisms are at preventing mutations and protecting us from rogue proteins, genetic mutations can occur anywhere. Because so many genetic disorders kill individuals before they can reproduce or even prevent reproduction, they are often very rare. This means that they can escape routine medical scrutiny. Developing genetic screenings for these rare disorders is an enormous task because of the sheer number of possible mutations. After all, as we've seen, disrupting even the simplest of functions can cause dramatic, life-altering symptoms.

Many rare genetic disorders arise from mutations in proteins involved in metabolism, the breaking down of molecules inside the body. Normal metabolism is complex, involving interwoven pathways that convert the food we consume into the energy and building blocks our bodies need to function. Proteins, in the form of enzymes, play a crucial role in catalyzing these metabolic reactions. Imagine a substrate, maybe a sugar or amino acid, that needs to be broken down into different parts by the countless enzymes in our bodies. One of those parts may be used to make another form of energy, and the other may just be waste that will get broken down further, recycled, or even sent out through our urine or feces. Because of the complexity of the metabolic process, genetic disorders can arise in a myriad of ways. Some lead to energy deficits, causing symptoms such as fatigue, weakness, and poor growth. Others result in the accumulation of toxic byproducts that affect specific organs or tissues.

In the 1970s, one example of a rare genetic disease rose to national prominence. A boy named David Vetter was born with severe combined immunodeficiency disorder (SCID). David, who became known as "Bubble Boy," spent his life confined to rooms surrounded by protective glass and plastic.[22] With no functioning immune system, any exposure to the outside world could kill him. At the time, there was little else that could be done other than to protect David from the outside world. NASA even built him a special suit so he could venture outside and interact with people face to face, even if behind a clear face mask. He died at age twelve after a failed treatment, a bone marrow transplant from his older sister. Today, only a hundred babies a year are born with SCID in the United States, and only one in a million worldwide.[23] While some of these children are born with mutations in proteins directly related to producing functional immune cells, many instead have a mutation in an enzyme called adenosine deaminase (ADA). This protein does something seemingly inconsequential: it grabs a single adenosine molecule, one of the four building blocks of DNA, and removes a small cluster of nitrogens and hydrogens, replacing them with oxygen.

When ADA doesn't function properly because of a mistake in its amino acid sequence, it can't properly remove nitrogens and hydrogens. This causes the buildup of a version of adenosine, called deoxyadenosine, that never replaces its nitrogen with oxygen. As this deoxyadenosine accumulates, it inhibits a key enzyme and essentially stops the immune system from functioning.[24] Specifically, deoxyadenosine stops any new DNA from being made, which in turn prevents new immune cells from being made. Immune cells replicate more than nearly any other cell in the body, and when that process is stopped by these toxic molecules, everything shuts down. There is no immune system without the ability to make more immune cells. The body is no longer able to fight invaders or protect itself from damaging molecules. Like other genetic diseases, a number of mutations can disrupt ADA's function. Some mutations decrease the activity of ADA, allowing it to still carry out its chemical reaction, but at a much less robust level. Other mutations completely obliterate its function.

ADA deficiency and its link to SCID underscores the intimate connection between genetic mutations, protein function, and overall health. Metabolic genetic disorders teach us that our bodies are delicate biochemical symphonies in which the harmonious interplay of proteins determines an organism's ability to survive. Each enzyme, each protein, is a crucial note in this symphony, and when even one is out of tune, the repercussions can resonate throughout the entire orchestra. Yet there is hope. Scientists are inching ever closer to bringing harmony back to the intricate metabolic melodies that are disrupted by genetic disorders. Recent research into these genetic disorders, including those that may arise spontaneously, like Alzheimer's, or heritable ones, such as cystic fibrosis, has opened the door to an entirely new, personalized approach to both treatment and prevention of these diseases. Advances in technology over the past several decades have allowed doctors to offer individual patients genetic screening that can predict which mutations in DNA will lead to malfunctioning proteins associated with disease. This has also paved the way for gene therapies, including with tools developed through CRISPR.

A NEW ERA FOR MEDICINE

Although CRISPR might be the most well-known tool for gene editing today, researchers have been treating diseases with other types of gene therapy for years. As early as 1990, gene therapy was used to treat

Ashanthi DeSilva, a young girl suffering from SCID because of a mutated adenosine deaminase enzyme.[25] Common ailments of healthy children, such as ear infections, common colds, and stomach bugs, plagued Ashanthi. Without a functioning immune system, she was unable to bounce back like most children. This kept her isolated in her home, away from the normal life of school, playdates with friends, and exploring the world around her. Ashanthi was dying and her parents were desperate to keep her alive. So when French Anderson and his colleagues, researchers from the National Institutes of Health, approached the DeSilva family with a new experimental treatment, they agreed. They were out of options and Anderson was confident that a new gene therapy treatment could work.

By this point, researchers had already succeeded in taking genetic material and inserting it into bacteria and plants, allowing these organisms to make proteins completely foreign to them. This was accomplished by using viruses that had been engineered to carry the new DNA and "infect" an organism. Essentially, researchers had hijacked the ability of a virus to insert itself into DNA and replicate.[26] Anderson believed this same approach could be used for treating patients with genetic disorders, including Ashanthi. He proposed that a modified virus carrying the DNA to make a normal functioning ADA enzyme could be injected into Ashanthi to correct the ADA sequence in her DNA. And he was right. Ashanthi's own molecular protein-making machinery, now able to produce a fully functional adenosine deaminase, took over the rest of the job of making Ashanthi well. After six months, Ashanthi's immune system had returned to normal, and within two years she was able to venture out of her house, return to school, and begin to live a normal life, free of symptoms. Her treatment, the first successful clinical trial for gene therapy, launched an entire new world of medicine into the spotlight. Since then, gene therapy has been used to treat other genetic disorders, cancers, and even infectious diseases like HIV and malaria.[27] While many of these treatments remain in clinical trials, it is clear that gene therapy and precision medicine will be part of the future of human healthcare.

In 2023, the first CRISPR-based gene therapy was approved by the Food and Drug Administration (FDA) for sickle cell anemia, a genetic disorder that causes an errant hemoglobin protein to alter the shape of red blood cells and impair their function.[28] This announcement made national headlines and even appeared in a sketch on *Saturday Night Live*. Clearly, gene editing has made its way out of technical research labs and clinical trials and into popular culture. But editing a person's DNA using CRISPR is still a new and controversial tool. As we harness the power of gene editing, researchers, and all of us, have to grapple with the ethics of genetically modifying humans. Consider the case of Chinese scientist He Jiankui.[29] Driven by the allure of groundbreaking genetic engineering, in 2018 He announced to the world the birth of twin girls whose DNA he claimed to have modified using CRISPR technology, making them supposedly immune to HIV. This bold step ignited a firestorm of controversy, sparking discussions about the ethics of altering the very fabric of human life.[30] The ability to play god and selectively edit or enhance certain genetic traits raises difficult questions that resonate deeply within society. The concept of designer babies conjures images of a dystopian future in which genetic privilege reigns supreme. How far should we go in manipulating the genetic dice? How long can humans cheat death? And who should have access to the most promising of these gene therapies?

My Grandpa died from Alzheimer's. I have a picture of him with me as a baby. It's one of my favorites. My uncle is on the living room floor napping, and my Grandpa is holding me on an old brown floral couch. Nobody looks ready to have their picture taken, but nonetheless, the camera captures everything about my Grandpa and our time together. It was easy and casual. He was a quiet man who always loved to tell about his time on the railroad. I would tell him about my boy crushes, and he would always respond in his subtle Southern accent, "Well, isn't that nice, darlin'," and give me a big squeeze.

My Grandpa ran a car mechanic shop out of the garage—Wright Electric—so he could be close to his family. I was always proud to see my last name on a sign. When we visited, I would pick up colorful cut wires in the driveway on my way up to the front door and keep them in my pocket to admire on the way home the next day. My Grandpa would come in through the back door all covered in oil and sweat when it was time to eat. With his strong, broad chest, his white shirt and jeans, and the rugged lines etched into his face from years of hard work, he looked like he could have walked out of a James Dean movie. And he called us all "darlin'" as we fought over who would get to sit next to him.

I was able to visit a few years before he died. My parents, sister, aunts and uncles, and a few cousins all made the trip to Anderson, Indiana. I arrived with my parents and sister at the tiny house on School Street before everyone else. I quickly went to be with him. He was in a rocking chair in the middle of the room with his favorite music, old choral hymns, playing loudly in a speaker set my aunt and dad had gotten for him. He smiled when I walked in. "Well, looky there, such a fine lady," he said. I sat down across from him and asked him how he was doing. He told me he liked his music. I showed him pictures of my cats and commented on his new sweatpants. We sat together for a long time as everyone trickled in. He smiled. But he didn't know us. All he said to me was, "We are connected, we all belong to each other." I sat next to him at lunch, cutting up his food. Helping him bring a fork to his mouth. I snuck him the extra piece of cake he asked for. I knew it was likely the last time I would be with him like this. Sitting at the dining table, the door to his garage right next to us. His favorite hymns playing in the background. A granddaughter feeding her grandfather. Later, when it was about time to leave, I kneeled down next to where he sat in his chair, his worn Bible on the table next to him. I held his hand and asked him if he remembered me. No response. "I am Maggie, your granddaughter. Barry's daughter. And I love you so much." In a tiny bit of clarity, he smiled again and laughed, a bit embarrassed. "Of course, darlin.' I knew it but I didn't know it. And I love you too."

Watching someone you love lose their sense of self and connection to others due to a natural bodily process gone wrong, as happens in

Alzheimer's, is devastating. And despite the promise that gene therapy brings for treating and potentially curing these disorders, most protein-based diseases continue to elude scientists. Almost twenty years after my grandmother's death from ALS, there is still no change in the outcome for others: the disease is a death sentence for all who are diagnosed. Those with Alzheimer's still fade away from their memories. Proteins still misfold. Mistakes are made in our DNA that cannot be repaired.

Our bodies will stop functioning one way or another. We cannot cheat death. But as we learn more and more about the great symphony of proteins that make us who we are, we can better understand ourselves and the world around us. And with that knowledge comes the ability to create new notes—to design something that nature has never seen before. For the hope of gene editing doesn't end with correcting mistakes, but rather with doing the miraculous: inventing entirely new proteins.

10

RESURRECTION

SHAHIR S. RIZK

MANY OF MY EARLIEST memories take place in a small church in my childhood hometown of Zagazig, Egypt. Murals of Noah's ark or Moses parting the Red Sea were painted across the walls, the first of many seemingly impossible stories I remember hearing from the Bible. We sang songs about mighty battles where the sun stood still in the sky for hours and city walls tumbled instantly. We listened to accounts of burning bushes, sticks turning to snakes, talking donkeys, and a man being swallowed by a whale and surviving for three days—fantastical tales of a natural world that was miraculously transformed and defied all rules. And as small children, dressed in our Sunday best, we prayed for miracles of our own.

In the sanctuary, stained-glass windows on either side of the room broke the light into rainbows, decorating the pulpit every Sunday morning and again in the evening. We sat for what felt like hours, listening to the preacher and singing hymns about death and resurrection. I can still picture it now, emerald greens and maroons in the pattern of palm leaves, a representation of Christ riding a donkey on Palm Sunday. The people covered the streets with palm branches for their Messiah. One week later, he was dead, crucified. But he came back to life and rolled away the stone covering his tomb. The ultimate miracle—defeating death.

Growing up in the church meant my early life was defined by a string of miraculous stories. Factual or not, they meant the world around me was magical too. All things are possible, I was told. At any moment I could witness a miracle, unexplainable by the rules of nature as I knew them. By the time I was an adult, the stories of my childhood didn't hold the same power, but they had laid the foundation for looking at the world through a different lens. There was something miraculous about even existing, taking a breath in and out, my heart beating at its own rhythm without my even telling it to. And I was surrounded by other living, breathing, heart-beating creatures.

Many people would argue that religious teachings of miracles and resurrections have no place in the world of science. Perhaps there is some truth to that. But without the ability to imagine the world as brimming with possibilities we haven't dreamed up yet, scientific discovery would grind to a halt. Without the magic of daring to take on the biggest of questions, countless scientific discoveries would never happen. The stories of death and resurrection at the heart of religious texts have inspired so much science. Being able to explain how the world has become as diverse and adaptable as it is doesn't take away from the astounding realization that we are surrounded by a vast, rich world—something that in this life, none of us will be able to fully grasp. What could be more miraculous than being alive, here, now?

If you Google the word "evolution," you will most likely find an image of a series of silhouettes. On one side is a chimpanzee hunched over, resting on all fours; on the opposite side is a muscular human male, standing upright. Between those two silhouettes is a series of changes showing an incremental transition between chimpanzee and human. This image is not only misleading; it promotes one of the most common misconceptions about the process of evolution. It suggests that over time, a chimpanzee became human or that that somehow, an organism can keep adding features in a linear progression to achieve more and more complexity. It also suggests a "missing link"—some form of animal that is between chimpanzee and human. This way of thinking dominated scientific thought in the nineteenth century, when Darwin's theory of speciation by means of natural selection was first introduced. In some cases, these false ideas were used to justify the oppression, colonization, and systematic eradication of indigenous populations in Africa, Asia, Australia, and the Americas, who were viewed as less evolved.[1] Unfortunately, this image still dominates popular beliefs about the theory of evolution. But in fact, the process is very different.

Instead of a linear progression, the evolution of life is better imagined as a giant tree full of branches, with each branch containing many leaves. Each year, the tree expands, extending more branches and sprouting new leaves. Each leaf is a living organism, dwelling in its environment on the branch where it first sprouted. Each organism, represented by a single leaf on the tree, is related to all the other leaves on the tree. The leaves on the same branch are more closely related to each other than are the leaves that live on branches far away from each other. Chimpanzees and humans, for example, are two leaves living on the same branch, very close to each other. In that sense, we cannot say that humans came from chimpanzees, just as we cannot say that one leaf came from another leaf. We can say that humans and chimpanzees share a branching point, the same way two leaves share the same branch. In evolutionary terms, that branching point is known as a common ancestor, an organism that was neither chimpanzee nor human, but an ancestor of both. Tracing the tree branches back toward the stem, we can find the point at which any two leaves are connected, the point where the branches first split from the main trunk. Similarly, we can go backward and find a common ancestor for both humans and dogs. Reaching even farther back into the tree, we can find a different branching point, an earlier common ancestor of both humans and bananas. Finally, we can reach all the way back to the tree trunk to find our earliest ancestor, an ancient unicellular organism named LUCA, the "last universal common ancestor."[2]

Long before humans discovered the molecular basis of evolution or how genetic information in DNA directs protein function, people all over the world used the cycle of reproduction, growth, and pruning to drive the evolution of countless plants and animals. Farmers crossed different strains of plants to get diverse offspring, each with a unique genetic makeup. Then, by selectively keeping the offspring with desired characteristics, they pruned the population, allowing only select members of the offspring to propagate their genes to the next generation. The results were bigger ears of corn, wheat that could withstand cold snaps, and more nutritious potatoes. Animal husbandry has also relied on this cycle of selective breeding and culling to drive the accumulation of preferred traits in livestock, or even to create entirely new species not previously found in the wild, like cows and dogs. In the past few decades, with new advances in molecular biology and a deeper understanding of the relationship between genes and proteins, this type of directed evolution has been brought into the lab. The cycle of mutation and selection is being used to drive the evolution of proteins with brand-new functions and structures.

CONQUERING DEATH

In the complex landscape of the human body's defenses, a fierce skirmish can break out—a battle often unseen yet profoundly felt by those who endure it. It's a clash within, where the immune system turns against the body, attacking the very tissues it is meant to protect. When it comes to rheumatoid arthritis, joints become battlegrounds. In this autoimmune disorder, the immune system mistakenly identifies the body's own tissues as invaders and launches relentless assaults on the synovium, the protective lining around joints. With no cure, living with the disease means enduring constant joint pain and inflammation, which often leads to disfigurement from the eventual destruction of cartilage and bone.

At the heart of this war lies a key player: tumor necrosis factor (TNF), a protein central to the body's inflammatory response.[3] Under normal conditions, TNF is dormant, activated only when a foreign invader launches an attack on our bodies. Once active, TNF triggers a cascade of inflammatory reactions meant to give the invading bacteria or virus an obstacle course. This defensive maneuvering takes a toll on the body. The inflammatory response triggered by TNF causes fevers, redness, swelling, and pain, all signs of an immune defense system launching an arsenal of weapons against the foreign organism. Once the threat is gone, TNF returns to its barracks, where it waits to be activated again by an external threat. But in rheumatoid arthritis, as in most autoimmune diseases, TNF becomes activated without an infection. This triggers excessive inflammation, causing the immune system to go haywire and driving the arsenal of immune weapons to turn against our own tissue.

Patients with rheumatoid arthritis can suffer for decades. Daily doses of aspirin and other anti-inflammatory drugs can offer some pain relief. Other more potent steroid-based drugs alleviate more severe symptoms but come with their own unpleasant side effects: digestive problems in the short-term, and debilitating muscle weakness, poor wound healing, and high blood pressure in the longer term. About two decades ago, however, a drug revolutionized the treatment of this painful, crippling disease. A twice-a-year injection of this drug is enough to relieve the agonizing symptoms many patients experience and to slow down joint degeneration. The drug, called adalimumab, belongs to a class of medicines known as biologics. More commonly known by its brand name Humira, this drug, like many biologics, is a protein.[4] Humira's job is to find TNF and block its function, ultimately suppressing the rogue inflammatory response associated with this debilitating disease. The development of Humira is a testament to how proteins can be engineered to precisely target a specific process in our bodies. Like many biologics, Humira was developed through the process of evolution. But unlike natural proteins, which evolve new functions over millennia, Humira took only a few years of research to develop. This powerful drug was intentionally engineered in the lab by directed evolution.

Humira is an antibody, the same kind of protein our B cells make to defend against foreign invaders. In mammals, antibodies must adapt to any infection they encounter, and in fact, they undergo their own process of evolution every time they are challenged by a bacterium or a virus. While evolution often takes centuries or even millennia to produce new species, the evolution of proteins can happen on a far shorter time scale. Recall from Chapter 7 that when there is an infection, B lymphocytes (B cells) spring into action as the main producers of antibodies. There are billions of B cells in our bodies.[5] Each cell is unique, displaying only one type of antibody on its surface. Each antibody waits for an interloper protein, known as an antigen, to appear. If one shows up, the helper T cells nearby must first confirm that it is indeed a foreign antigen before the attack begins. Any B cells that display an antibody with a good fit for the antigen will begin to divide frantically, making more and more copies of themselves. But while this replication frenzy is taking place in the activated B cells, a protein known as cytidine deaminase will intentionally produce mistakes in the gene coding for the antibody, particularly in the regions responsible for binding to the invader protein.[6] It may seem counterintuitive to make mutations in the binding region of an antibody that already recognizes the invader. After all, why break something that works? But the job of cytidine deaminase is to introduce controlled mistakes at a very high rate, fulfilling the first requirement for evolution—variation. This process, termed "hypermutation," creates hundreds of thousands of new variants of the original antibody, each a slightly modified version of its predecessor.[7] Inevitably, many of these modified antibodies will be unable to bind to the antigen. But this is a cost the immune system is willing to pay. The payoff is new variants that may have much higher affinity for the antigen, and so can be more potent inhibitors of the infection.

With this vast number of new cells, each with a new variant of the parent antibody, comes the second component of evolution—selection. In this stage, phagocytes swallow the antigen protein and display it on their surface, dangling pieces of it like bait. The B cells with the highest affinity for this bait will latch on, while those with low or no affinity will wash away into the vast lymphatic system. B cells that latch onto the antigen will receive a signal from neighboring T cells to proliferate and divide, creating more of themselves. The cycle of hypermutation followed by selection ensures that antibodies with the highest affinity are made in order to snuff out the infection as quickly as possible. This same strategy has been used to produce therapeutic antibodies with the goal of treating numerous diseases.

One of the earliest examples of therapeutic antibodies is trastuzumab, also known as Herceptin. After its approval in 1998, it quickly became widely used for treating breast cancer. Herceptin targets a protein on the surface of cancer cells known as the HER2 receptor.[8] This receptor mediates signals from growth factors and hormones, two types of proteins that can promote cell division. Cancer cells often have many copies of the HER2 receptor. This causes them to replicate uncontrollably and ultimately leads to tumor formation. Herceptin is an engineered antibody that can bind to the HER2 receptor with high affinity and disable its function, slowing the growth of the tumor and prolonging the lives of countless women who suffer from the disease. Roughly 85 percent of women treated with Herceptin, in combination with chemotherapy, experience at least a ten-year survival rate.[9]

Herceptin was developed using the same immune response we use to produce antibodies against foreign invaders. In fact, to obtain Herceptin, lab mice were injected with the human HER2 receptor protein. Believing that the receptor was a foreign invader, the mice began to make antibodies against it. Researchers at Genentech isolated the antibodies from the mice and painstakingly investigated each one for its ability to block HER2. But the work was not quite done. Injecting an antibody from a mouse into a human comes with its own harmful side effects. The human body will recognize the mouse antibody as a foreign protein and immediately destroy it. So the researchers began to systematically make mutations in the mouse antibody to convert it into a "humanized" antibody, one that would not be recognized as foreign by the human immune system. After a series of successful clinical trials, this humanized antibody became one of the first examples of engineered proteins used as targeted therapies.

Not long after the development of Herceptin, it became clear that directed evolution of antibodies would be an effective way to develop novel therapeutic antibodies. Scientists working at Genentech, the same company that developed Herceptin, began expanding the technique. As more became known about the roles of certain proteins in the formation and spread of tumors, these proteins became obvious targets for this type of antibody-based therapy. One protein in particular, known as vascular endothelial growth factor (VEGF), came to the forefront of cancer research in the late 1980s, when it was shown to be directly involved in the spread of several cancers. Like the HER2 receptor, the VEGF receptor responds to the growth factor VEGF to stimulate the formation of blood vessels, providing more oxygen to growing tissue. Led by Napoleone Ferrara, a team at Genentech showed that overactive VEGF is a key factor in the growth and proliferation of cancer cells because solid tumors can't grow beyond a certain size without a sufficient blood supply. When cancer cells overproduce VEGF, it induces new blood-vessel growth in the tumor, enabling the cells to grow further and increasing the likelihood that they metastasize.[10] By blocking the action of VEGF, Ferrara and his team hoped to slow down the growth and spread of tumors.

The team began by immunizing mice against the human version of the VEGF protein. Using the mouse immune system gave the researchers lots of antibodies against VEGF, each with the potential to block its cancer-promoting activities. The researchers isolated the B cells, then separated them to determine which ones were producing anti-VEGF antibodies. Then they tested each of those antibodies to see which of them would stop cells harboring VEGF from growing. This was like finding a needle in a haystack, but in 1993, the team found one antibody that could inhibit the growth of tumors in mice. While the scientists were encouraged, they still had to turn the mouse antibody into a humanized antibody for use in humans. They accomplished this by transplanting the parts of the antibody responsible for binding to VEGF onto a human antibody scaffold. Finally, in 2004, the humanized antibody targeting VEGF, bevacizumab—now known as Avastin—was approved by the FDA for treating colorectal cancers. The success of Avastin in clinical trials was unprecedented. It increased the survival of patients, most of whom were terminally ill, by more than five months. This was one of the largest increases in survival ever seen in an FDA Phase III trial for colorectal cancers.[11] Avastin is now used to treat a wide variety of cancers, including brain tumors, metastatic cancers of the bowel, and lung and cervical cancers.

The success of both Herceptin and Avastin, two of the earliest antibody-based therapeutics, cemented the role of protein engineering in the development of new therapies. What makes antibody-based therapy particularly effective is its high level of specificity. Traditional chemotherapy usually relies on stopping the growth of fast-dividing cells. While this helps slow down the growth of tumors, it comes with severe side effects. Chemotherapy does not discriminate between tumor cells and healthy ones in a patient, and as a result, the body suffers greatly. The patient experiences hair loss as hair follicles become damaged. Immune cells, which typically divide at high rates, die out, making patients more susceptible to infections. On top of it all, persistent nausea, vomiting, and digestive problems occur because chemotherapy affects the fast-dividing cells lining the digestive system. But the majority of antibody-based therapies work very differently. They rely on the high specificity of antibodies to ensure that only the intended target is harmed, avoiding the prevalent side effects seen with chemotherapy. In the battle against cancers, if chemotherapeutics are indiscriminate carpet bombs, proteins are computer-guided missiles that nimbly avoid collateral damage.

HEALER PHAGE

Even with the wild success of Herceptin and Avastin, the process of using the mouse immune system to generate antibodies against cancer targets is no small feat. After identifying a suitable mouse antibody, the protein must be humanized to avoid rejection by a patient's own immune system. Not to mention that using mice to isolate antibodies requires specialized animal facilities that must adhere to strict codes of humane treatment. Such challenges are significant obstacles in the development of new antibody-based therapies. As a result, scientists began looking for ways to make the process more efficient. This is when a different approach came to the scene, a new method of directed evolution known as phage display.[12] This powerful technique eliminates the need to humanize the resulting antibody. Like all types of directed evolution, phage display relies on a cycle of the same two components: mutation and selection. But instead of relying on the B cells of the mouse immune system to generate and display a wide range of antibodies on their surface, phage display uses tiny viruses that infect bacteria. These phage particles are harmless to humans but can serve the same function as B cells in the process of making therapeutic antibodies.

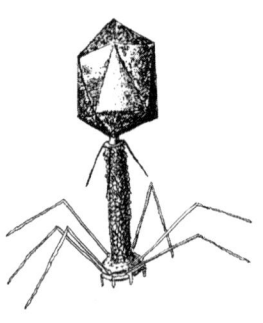

To perform phage display, first the gene that codes for a human antibody is fused to the genetic code of a small phage. When done precisely, the resulting phage will display the antibody on its surface. Then mutations in the regions responsible for binding to an antigen are made in the antibody scaffold, producing up to 100 billion new phage particles, each one displaying a unique antibody. This immense number of new antibodies is known as a library. Each member of the library is a slightly different variation of the parent antibody. Because phages are so tiny, the entire library can fit in a volume about the size of a drop of water. Then comes the second part: selection. This is where the scientists need to isolate one or a handful of antibodies with the desired property, namely, the ability to bind to a target protein. During this stage, the entire library of billions of phage particles is exposed to a target protein immobilized on a plastic surface. Phage particles displaying antibodies with the ability to bind to the target will stick, and those unable to bind to the target will simply be washed away. In this manner, the scientists are

Phage display is a powerful technique used to engineer proteins with new binding capabilities. A phage particle, like this T4 phage, can be used to screen through billions of mutants (variants) of a protein to identify those that can bind to a target protein. Phage display has been used to make a number of therapeutic antibodies, including Humira, a biologic used to treat a variety of autoimmune diseases.

able to select for new antibodies with the ability to bind precisely to a target protein and remove the antibodies that do not bind to the target.

Phage display was first developed in 1985 by George Smith and Gregory Winter. In the initial experiments, only small proteins were displayed on the phage particles. Then eventually, antibodies were displayed. Library sizes grew bigger and more diverse as new biotechnology techniques became available. The technology has had a huge impact, and Smith and Winter were awarded the Nobel Prize in Chemistry in 2018 for developing it.[13] The technique has been widely used to develop a number of newly engineered proteins with several clinically approved therapeutic antibodies.[14] In fact, Humira, the antibody used to treat rheumatoid arthritis, was engineered using phage display. In 2002, it became the first FDA-approved antibody made by phage display for the treatment of human diseases. Before long, it became the most widely prescribed antibody-based drug.

There are many advantages to using phage display in directed evolution. First, no animals are needed to make the antibody library or for any part of the selection process, because the phage particles replace the use of animal B cells. This makes the process much easier and more cost effective. Second, unlike traditional antibody generation, which requires the mouse immune system, phage display can eliminate the challenging step of humanizing mouse antibodies by starting with human antibodies. Third, since the entire process can be done in a test tube, the conditions for phage-display selections are not limited to the bloodstream of an animal. Parameters like pH, temperature, or salt concentrations can be fine-tuned to achieve the ideal conditions for finding an antibody with the desired properties. Phage display has been carried out in the presence of detergents found in shampoo to produce antibodies that can prevent dandruff.[15]

Antibodies have become highly effective tools in our fight against disease. Our naturally made antibodies protect us from pathogens, and the more recently engineered antibodies now fight against the spread of cancers and the progression of autoimmune diseases. These processes are rooted in the natural ability of antibodies to bind selectively to a target protein and block its function. Recall from Chapter 7 that the key to how an antibody works is an exquisite shape complementarity that allows the antibody and the antigen to fit together like puzzle pieces. As a result, proteins that are overactivated in disease, such as TNF, HER2, and VEGF, can be subdued by highly selective antibodies. But often diseases are not caused by overactive proteins; instead, they are the result of a protein not being active enough. When a protein is not working at a sufficient pace, it cannot keep up with the demands of the body. This protein "loss of function" is extremely common in genetic disorders, and in some cases, an antibody has been considered as a way to reactivate the function of the protein in order to bring it back to life. This task is much more challenging. After all, it is easier to muck something up than to fix it. But recent work in our own lab at Indiana University, South Bend has shown that phage display can be used to produce a new kind of activator drug—antibodies designed to rescue the function of a failing protein.

One form of brain cancer, known as astrocytoma, is a particularly aggressive disease. Once diagnosed, fewer than 10 percent of patients survive more than five years. In some cases, astrocytoma is characterized by a unique mutation in a key metabolic enzyme: isocitrate dehydrogenase (IDH). This enzyme is responsible for mediating an important step in the breakdown of carbohydrates, fats, and proteins for energy production. But a mutation in IDH changes the 3D structure of the protein in a significant way, preventing it from working properly.

Using phage display, our group, in collaboration with scientists at the University of Chicago, isolated an engineered antibody that binds selectively to the healthy structure of IDH. When added to the mutant enzyme, the antibody could bring the enzyme back to life by forcing it to adopt the correct 3D shape required for it to function.[16] While this success has been demonstrated only in the lab, it provides hope that a way can be found for this sort of rescue to happen *in vivo*, in human patients. With an estimated seven thousand rare genetic disorders, most of which are a result of a faulty enzyme, this approach offers a glimpse into a future where activator drugs could be tailor-made for the treatment of innumerable diseases.

Directed evolution has clearly shown great success in engineering antibodies with new functions, some of which are revolutionizing how we manage disease. But the use of directed evolution is not limited to engineering antibodies. In fact, virtually any protein scaffold can be used. Like natural evolution, variation followed by selection is the name of the game. With twenty natural amino acids to choose from at any given position within a protein, the possibilities for engineering new proteins are endless. Consider a tiny protein made up of five amino acids. Since each of the five positions can be any of the twenty amino acids, the number of possible proteins is $20 \times 20 \times 20 \times 20 \times 20$, or 20^5. This means that there are 3.2 million different combinations of amino acid sequences that can be used to make a protein only five amino acids long. Of course, most proteins are far longer than five amino acids. Even a small protein like insulin contains 51 amino acids, while hemoglobin has 574 amino acids. If we take a protein that is 100 amino acids long—not very long by most protein standards—the number of different combinations of amino acids would amount to 20^{100} or 1×10^{130}. That's 1 with 130 zeros! This is an unimaginably large number. In fact, there's not enough matter in the entire observable universe to make every single amino acid combination.[17] This tells us that even with the vast diversity of proteins in all of the animals, plants, fungi, and bacteria that exist today, and have ever existed, nature has explored only a minuscule proportion of the potential amino acid sequence combinations— only a snowflake at the top of the amino-acid-sequence iceberg. This is good news for protein engineers. It means that there will be many, many new ways to string amino acids together in ways that nature has never seen before.

THE POWER OF ONES AND ZEROS

Some of us are old enough to remember the earliest ways we used to connect to the Internet. A computer was hooked up to a dial-up modem through a telephone line, making strange beeps and boops in order to communicate with the World Wide Web. Even a small, heavily pixelated image took minutes to load on the screen, and sending a short message was a delicate and unreliable affair, especially if someone in your home decided to pick up the phone, interrupting the entire process. But computers became faster. Processor speeds made computers more powerful, and wireless technology made communication lightning fast. With the exponential growth in computational power, scientists began to explore the use of computers in the design of proteins. After all, if the protein design problem could be reduced to a set of calculations, then computers, which are in essence giant calculators, should be able to help solve the problem much more quickly.

Initial work on computational protein design used computers to ask limited questions, mostly modeling how mutations could affect the structure of a protein. But in 1997, Stephen Mayo's group at Caltech reported the first protein completely designed by a computer.[18] The new protein, named FSD1 (short for "full sequence design 1"), was just twenty-eight amino acids long. Mayo was one of the early pioneers of computational protein design. At the time FSD1 was made, he was the only Black faculty member at Caltech.[19] Later, he would become the first Black professor to earn tenure there.[20]

Mayo wanted to answer a big question: could a fully folded protein be made entirely by computational methods? He began with the backbone structure of a protein called a zinc finger. The protein is composed of a single alpha helix and two beta strands surrounding one zinc ion. Mayo hoped to strip the original zinc finger protein of its amino acids and to replace them with new ones without disrupting the overall structure. This is a challenging endeavor considering that there are over two trillion trillion trillion different combinations of amino acids that could be put together to create the twenty-eight amino acid–long protein chain. With the computational power at the time, it was nearly impossible to try every single combination of amino acids and systematically determine which ones to test. So instead, Mayo came up with a clever solution to scan through the possible solutions to this problem. He and his coauthor, Bassil Dahiyat, used a computer algorithm known as dead-end elimination (DEE). Rather than try to find the best answers to the problem, the algorithm searched for amino acids that could not possibly be part of the final solution and eliminated them. For example, if an amino acid was too large to fit at a specific position, then the algorithm would reach a dead end, concluding that that specific amino acid chain would not work. The process was repeated over and over, pruning the wrong answers, gradually simplifying the problem as wrong answers were eliminated, and ultimately leading to a reasonable solution. Using the DEE algorithm, Mayo and Dahiyat came up with a new sequence that was different from the original zinc finger but had an identical fold. The work became a milestone in computational protein engineering and marked the beginning of a now flourishing field.

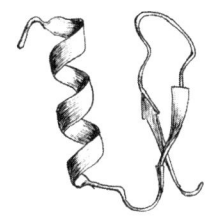

The first protein to be designed completely by a computer program was FSD1. The small protein, composed of one alpha helix and two beta strands, is based on the structure of a zinc finger fold.

After Mayo published his work, researchers began to improve the computational algorithms and use smaller and faster machines as the technology raced forward with better hardware and software. In 2003, David Baker's group at the University of Washington described the design of a new protein fold, a structure never seen before in nature. The protein was ninety-three amino acids long and combined alpha helix and beta sheets in a new topology that they named Top7.[21] Baker's group continued to show resounding success in the field of computational protein design. In 2008, the group reported the design of a brand-new enzyme using computational methods.[22] This was an accomplishment the entire field had been anticipating, and the publication became an instant hit, launching Baker to the forefront of computational protein design. The enzyme reported by the group was shown to carry out a reaction known as Kemp elimination, whereby the protein breaks a bond between a nitrogen and an oxygen in a small organic molecule. The reaction by itself is slow, but with the newly engineered enzyme, the reaction goes a hundred thousand times faster. The group then used directed evolution to improve the designed protein, increasing its activity by an additional ten-fold. That same year, the group repeated their success by reporting a new set of computationally designed enzymes.[23] This time the enzymes were designed to catalyze a retro-aldol reaction, which is often used in carbohydrate metabolism. The group tested seventy-two different designed enzymes, of which thirty-two were able to carry out the reaction. This represented a massive success rate. The group also demonstrated that their designed enzymes could perform the reaction using four different sets of amino acid sequences. Most importantly, the structures of the newly designed proteins were nearly identical to what the computational algorithms had predicted. Using new enzymes to carry out these reactions provides an eco-friendly way to replace traditional methods, which usually require harsh or dangerous chemicals.

David Baker has continued to demonstrate novel ways in which computational algorithms can be used to design proteins with new structures and functions. In 2018 his group designed its own version of the green fluorescent protein found in jellyfish.[24] Unlike GFP, the designed protein could not produce fluorescence on its own. Instead, its beta-barrel structure was designed to surround a fluorescent dye similar to the fluorescent core of the naturally occurring GFP. This novel protein represented a new level of success for computational protein engineering techniques, offering ways to fine-tune and even improve the fluorescent properties of proteins found in nature. The ability to continually engineer new types of proteins and raise the bar for protein engineering requires a great deal of creativity and a network of dedicated scientists working together as a team.

Protein engineering also requires a massive amount of computational power. Due to the incredibly complex nature of proteins, the computational process requires multiple processors working together, each solving little bits of the immense number of calculations. Even with large computational facilities containing hundreds of processors, the demand for computational time is always high. Because of this, David Baker and his group have enlisted the help of the public in carrying out the intense protein design process.[25] Their application, called Rosetta@home, enables people all over the world to help solve protein-design problems.[26] Participants allow researchers to use their personal computers for protein-structure prediction calculations when the computers are otherwise not in use. With more than 1.3 million users on 250,000 active hosts worldwide working together, Baker's group has been able to crowdsource research on protein engineering. The group also developed Foldit, an online platform that has turned protein folding into a collaborative game accessible to players around the world.[27] Participants, whether experienced scientists or just curious gamers, engage in virtual challenges to predict the most stable and biologically functional structures for various proteins. This crowd-driven approach to protein folding has not only expanded the pool of contributors, advancing the fields of biotechnology and computational biology; it has also fostered interdisciplinary collaboration and public engagement in scientific research. But the Baker lab hasn't focused just on surmounting general protein design obstacles. They've also set their sights on developing novel therapeutics to fight some of the greatest health threats of our time.

In 2020, as the COVID-19 pandemic spread across the world, hospitals were overwhelmed with patients, and pharmaceutical companies raced to develop a vaccine. Many scientists shifted their attention to studying how the virus infects humans and how infections can be prevented or treated. In the Baker lab, members were focused on engineering proteins that could help in the battle against the virus. The key to the COVID-19 virus's ability to infect the human body is a protein on its surface known as a spike protein. This is how the virus latches on and gains entry to the inside of the cells of the respiratory tract, where it proliferates and causes disease. The Baker group wanted to block the spike protein from binding to receptors on the surface of human cells. The group designed a set of small proteins that were able to latch onto the spike protein, specifically the parts of the protein responsible for its invasive abilities.[28] The group used computational algorithms to build the tiny proteins with the right shape-complementarity for the COVID-19 spike protein, to try to mimic how an antibody can block a target protein. The researchers knew that the spike protein interacts with a receptor on our lungs known as ACE2. The algorithms started with a 3D model of the part of ACE2 that the spike protein latches onto, then began to build around it in the hope of making proteins that could disrupt the interactions between the spike protein and ACE2. The newly designed proteins latched onto the spike protein with high affinity and in a specific manner.[29] To test if the designed proteins could protect from a COVID-19 infection, the protein was delivered to mice through their nostrils before the mice were exposed to the virus. The results showed that the engineered proteins were sufficient to protect the mice from infection and were effective against several different strains of COVID-19, offering hope for their use as prophylactics in humans.

Over the past three decades, computational scientists have made great strides in the design of new proteins. Yet one major problem has persisted: how to predict the structure of a protein based on only its amino acid sequence. This is called "the protein folding problem." There is no model available today that can predict how a string of amino acids will fold into its functional form, or if it will fold at all. This may seem like a minor problem; after all, proteins naturally fold into precise 3D structures, and they do this in a fraction of a second. But for scientists creating a protein shape from a string of amino acids, knowing how this happens is essential. Even with a very short protein, there are nearly endless ways in which the different parts of the protein can come together as it wiggles and twists. The incredible flexibility of the bonds connecting atoms in the protein compound the problem. Twisting just one bond in the wrong way can have immense consequences for the way the entire protein folds.

For years, scientists grappled with this seemingly unsolvable problem, until AlphaFold came on the scene. The brainchild of Deep-Mind, the artificial intelligence powerhouse created by Google, AlphaFold quickly became a revolutionary force poised to transform our understanding of protein folding.[30] The turning point was during the Critical Assessment of Structure Prediction (CASP) competition—an Olympic-like event for biochemists and computational biologists striving to predict protein structures. Every year since 1994, CASP participants have raced to see who can most accurately predict protein structures when given only their amino-acid sequences.[31] The participants receive a score on a hundred-point scale reflecting how well their predicted structures match the experimentally determined protein structure. Throughout the history of the competition, the scores for the more challenging protein structures have hovered between forty and sixty, meaning that they did not match well with the actual structures of the proteins.

Using neural networks and other machine learning algorithms, AlphaFold began to decode the language of proteins and the complex symphony of interactions, forces, and shapes they adopt. The 2020 CASP competition was AlphaFold's moment, and as the competition unfolded, AlphaFold dazzled the scientific community with its results. While much of the world was still under quarantine, an announcement was made to the AlphaFold team over a Zoom meeting. The group had not only won the competition; they had also exceeded all expectations. The group's AI algorithms were able to predict the structure of several dozen proteins, given only their amino-acid sequences, with a median score of eighty-seven, a full twenty-five points above the next highest score. The structures generated by AlphaFold's AI algorithms were nearly identical to those determined by experimental methods. The groundbreaking accomplishment inspired protein scientists and biologists worldwide. It wasn't just a victory in a competition; it was a paradigm shift in our ability to decipher life's most fundamental building blocks.

AlphaFold's success lay in its exquisite ability to learn from the two hundred thousand or so known protein structures, finding patterns and deciphering the rules that ultimately govern how proteins fold. This is similar to how AI tools that generate images work. AlphaFold treats protein structures as any three-dimensional object. It learns the structural patterns from the library of available structures, assessing the propensity of certain sequences for forming alpha helices or beta sheets, then generates shapes from new sequences with unknown structures using what it has learned. But image learning is not the only feature of AlphaFold. Like the AI tools used to generate language, another component of AlphaFold treats the sequences of proteins as words or sentences, deciphering their patterns and assigning them structure the same way a text-generating AI tool assigns meaning to words.

Since the development of AlphaFold, many new tools involving AI have emerged, each with new capabilities for specific applications. Some tools allow scientists to see how two different proteins fit together, for example, how a receptor on a cell surface receives a signal from a hormone. Other tools can be used to quickly scan through millions of drug molecules and see which best fits a protein structure, providing insight into drug design. What's most impressive is that tools like AlphaFold are freely available for anyone to use. There is no need for large computational facilities with multiple parallel processors. Anyone can use a laptop or phone to plug a sequence into the AlphaFold online tool and, within minutes, get an incredibly accurate picture of what the structure probably looks like. This combination of AI and biochemistry puts proteins in the hands of more people, redefining who can participate and create. These new tools are democratizing scientific research by increasing accessibility and removing the traditional barriers to who can become a scientist.

The revolutionary work of David Baker and the team of scientists who developed AlphaFold did not go unnoticed. In 2024, the Nobel Prize for chemistry was awarded to David Baker for his efforts in using computers to design new proteins, and to Demis Hassabis and John Jumper of Google DeepMind for protein structure predictions. The prize recognized the power of AI to help us understand how a sequence of amino acids folds into a functional three-dimensional structure. It also underscored the incredible potential for designing new proteins that nature has never seen before.

ENGINEERING THE FUTURE

Beyond their importance for human health, engineered proteins may become our most powerful weapon against climate change. With the rising temperature of the planet, an increase in greenhouse gas emissions, and the huge amounts of pollutants and plastic waste produced every year, protein engineering may offer hope for a cleaner future. One promising application for protein engineering is the detection of pollutants. Scientists are now turning proteins into "biosensors" that can detect a wide range of molecules, especially those that negatively impact the environment.

Recently, our group at Indiana University engineered a bacterial protein to bind to glyphosate, the active ingredient in RoundUp, the most extensively used herbicide in the world. Each year, 300 million pounds of glyphosate are sprayed on crops, much of which ends up in groundwater, lakes, rivers, and streams. While initially considered harmless, glyphosate has been associated with negative health effects and was classified as a carcinogen by the state of California in 2017. Starting with a bacterial protein that binds to a naturally occurring molecule known as 2-aminoethyl phosphonate, we made mutations in the protein to change its composition, altering its function and turning it into a glyphosate-binding protein.[32] When a fluorescent molecule was attached to the engineered protein, the amount of light emitted by the fluorescent chemical indicated the presence of glyphosate in soil or water.

Protein engineering also offers a new avenue for carbon capture. Each year, coal-burning plants emit nearly nine billion metric tons of CO_2 into the atmosphere, adding to the problem of climate change and contributing to a warmer planet. Current technologies for capturing carbon dioxide as it travels through the chimneys at these plants require large amounts of energy. To tackle this problem, scientists recently sought the help of the enzyme carbonic anhydrase, which naturally combines CO_2 with water to make bicarbonate, known to most of us as baking soda. Carbonic anhydrase is one of the fastest enzymes on the planet, with a rate of about one million reactions per second.[33] Unfortunately, the enzyme is not tolerant of the harsh conditions inside the reaction center of a chimney. The enzyme must be able to withstand temperatures near the boiling point of water and harsh chemical conditions that would denature most proteins. To overcome these

challenges, scientists from the biotech firm Codexis used directed evolution to modify carbonic anhydrase with the goal of making a more
stable enzyme while preserving its impressively fast activity.[34] They
began by making mutations, producing roughly 27,000 different variants of the enzyme. They then meticulously challenged the variants
by putting them in alkaline conditions (high pH) until they had identi-

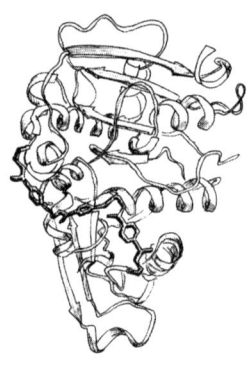

A model of an engineered
protein that can digest PET,
the main component in
many plastic products. With
nearly 200 million tons of
plastics accumulating
worldwide, this enzyme is
among many new engi-
neered proteins that holds
promise as a major weapon
against pollution.

fied enzymes that remained active. In the end,
the researchers were able to engineer a new
carbonic anhydrase that could work at a temperature of 107°C, while enhancing the rate of
CO_2 absorption by twenty-five-fold. In addition, the enzyme was able to withstand high
pH that would denature most proteins.

More recently, protein engineers have
turned their attention to tackling the problem
of plastic waste. Of the 380 million tons of plastics produced annually, around 10 percent is
dedicated to a material known as polyethylene
terephthalate (PET), which is used to make synthetic fabrics and plastic bottles. There is a
limit to how many times PET can be recycled
without losing its integrity, so each year, tons
of PET waste end up in landfills or find their
way into the oceans. More concerning is the
fact that some PET eventually turns into micro-
plastics, endangering marine ecosystems and
posing a potential threat to human health. In 2012, a group of scientists in Japan isolated an enzyme from leaf-branch compost using a
large-scale DNA sequencing approach.[35] The enzyme was found to digest
a number of fatty acids and was able to break down PET at a modest
rate. In 2020, scientists in France took a computational approach to
making changes to the PET-eating enzyme. They were ultimately able
to engineer a much more stable plastic-digesting enzyme that is
90 percent efficient in breaking down PET.[36] As little as one gram of

the engineered enzyme could break down over five hundred grams of PET within ten hours. Based on this new technology, a manufacturing plant for fully bio-recycled PET with the capacity to break down up to fifty thousand tons of PET annually is expected to be completed in 2025. The PET bio-recycling center will reside in France and will be the first of its kind.[37] In the future, engineered enzymes will digest the strong plastic polymer structure into smaller carbon-containing compounds that can then be extracted and used in a wide range of industrial applications.

Protein engineering is a beacon of hope in the battle against environmental pollution and climate change. As our planet faces rising temperatures, escalating greenhouse gas emissions, and a mounting tide of pollutants and plastic waste, engineered proteins offer innovative solutions for a cleaner, more sustainable future. In a few short decades, the immense advancements that have been made toward engineering new proteins stand as a testament to human ingenuity and innovation. One can only imagine what the future holds for protein engineering. It may quite literally save our planet.

EPILOGUE

ACCORDING TO THE BIBLE, the first miracle Jesus performed was turning water into wine at a wedding celebration. Later, he made a paste from dirt and spread it over the eyes of a man born blind. When the dirt was washed away, the blind man could see. The miracles became more and more elaborate—healing the sick and lame, calming storms, and raising a dead man from his tomb. These acts of supernatural power may seem far-fetched, defying our understanding of life. But every day, small miracles happen all around us. Within our liver cells, enzymes turn toxins into harmless molecules, defensins protect us from infections, and light receptors illuminate the world around us. And it all happens without us having to think about it. Inside every organism, proteins live in a world all their own—dynamic, responsive, alive. They are more than simple tiny molecular machines. In many ways, they are microscopic miracle workers. While our DNA may dictate who we are genetically, it is merely the instruction manual for how to put all the pieces together. Proteins, with their three-dimensional shapes, moveable parts, and complex chemistry, are what make us who we are. They remember the places we have been and where home is. They store the memory of citrus on our tongues during a cold winter long ago. They form the bond between a mother and child. They protect us from threats of all kinds. They connect us to the web of life on this planet—enabling us to taste the sweetness of a mulberry or witness the magical glow of fireflies. The enzymes inhabiting our microscopic cellular world extract energy from food, driving the movement of our muscles. The proteins of our immune system keep a constant watch, protecting us from foreign invaders. And in the end, proteins may be our own demise, finishing the story of our lives with death.

But we humans are not the only ones to benefit from proteins, nature's problem solvers. Trees quietly use proteins to capture carbon dioxide, turning this harmful greenhouse gas into a sweet nectar of maple syrup, and yeast enzymes perform a kind of miracle, turning grapes into wine. Photoreceptors inside the eyes of finches and robins use quantum physics to show the birds a color they call north, guiding their long migration journey in the sky. The saintly halo of a jellyfish is the glow of a fluorescent protein, and the flickering of fireflies is the enzymatic action of luciferase. And proteins in the blood of some wood frogs perform the ultimate miracle, a resurrection, allowing them to come back to life after being frozen solid.

With recent breakthroughs in protein engineering, new miracles are closer than ever before. Healing the sick with engineered antibodies that precisely target diseased cells is now routine. Biologics such as Herceptin, Avastin, and Humira are just three examples of more than 120 approved antibody-based therapies, and more flow through the drug development pipeline each year. Engineered proteins are now used to repair genetic disorders by acting as tiny molecular surgeons, extracting and replacing mutated genes with healthy ones. This type of targeted gene editing has shown great promise for the treatment of countless diseases. Moreover, protein engineering promises to quell the coming storm of global climate change. New enzymes are being engineered for biofuel production and carbon capture, and to degrade the millions of tons of plastic waste we produce each year. Other proteins are being designed to detect environmental pollutants and break down harmful chemicals.

Every day, the tiny machines we call proteins perform big miracles, and more are being developed to solve immense problems. The incredible diversity of proteins found in nature, combined with recent breakthroughs in protein engineering, foretell a time when designer proteins will accomplish tasks that far exceed what we can imagine today. Whatever our future holds, proteins are poised to lead the way to a healthier world and a more sustainable future.

NOTES

ORIGIN

1. Anirban Bhunia et al., "NMR Structure of Pardaxin, a Pore-Forming Antimicrobial Peptide, in Lipopolysaccharide Micelles: Mechanism of Outer Membrane Permeabilization," *Journal of Biological Chemistry* 285, no. 6 (Feb. 5, 2010), https://doi.org /10.1074/jbc.M109.065672.

2. Geoffrey Cooper, *The Cell: A Molecular Approach,* 2nd ed. (Sunderland, MA: Sinauer Associates, 2000).

3. "Gene," Talking Glossary, National Human Genomic Research Institute website, https://www.genome.gov/genetics-glossary/Gene.

4. Harold C. Urey, "On the Early Chemical History of the Earth and the Origin of Life," *Proceedings of the National Academy of Sciences USA* 38, no. 4 (Apr. 1952), https://doi.org/10 .1073/pnas.38.4.351.

5. Ashfaq M. Sheikh et al., "Elevated Levels of Serum Alpha(2) Macroglobulin in Wild Black Bears during Hibernation," *Biochimie* 85, no. 10 (Oct. 2003), https://doi.org/10.1016/s0300 -9084(03)00133-0.

6. Karen G. Welinder et al., "Biological Foundations of Health and Energy Conservation in Hibernating Free-Ranging Sub-Adult Brown Bear *Ursus arctos,*" *Journal of Biological Chemistry* 291, no. 43 (Oct. 21, 2016), https://www.ncbi.nlm.nih.gov/pmc/articles/ PMC5077189.

7. Hak J. Kim et al., "Marine Antifreeze Proteins: Structure, Function, and Application to Cryopreservation as a Potential Cryoprotectant," *Marine Drugs* 15, no. 2 (Jan. 27, 2017), https://doi.org/10.3390/md15020027.

8. Laurie A. Graham et al., "Hyperactive Antifreeze Protein from Beetles," *Nature* 388, no. 6644 (Aug. 21, 1997), https://doi.org/10.1038/41908.

9. Ravi Gupta and Renu Deswal, "Antifreeze Proteins Enable Plants to Survive in Freezing Conditions," *Journal of Biosciences* 39, no. 5 (Dec. 2014), https://doi.org/10.1007/s12038 -014-9468-2.

10. Patricio A. Muñoz et al., "Structure and Application of Antifreeze Proteins from Antarctic Bacteria," *Microbial Cell Factories* 16, no. 1 (Aug. 7, 2017), https://doi.org/10.1186/s12934-017-0737-2.

11. Aaron Hakim et al., "Crystal Structure of an Insect Antifreeze Protein and Its Implications for Ice Binding," *Journal of Biological Chemistry* 288, no. 17 (Apr. 26, 2013), https://doi.org/10.1074/jbc.M113.450973.

12. Muñoz et al., "Structure and Application of Antifreeze Proteins."

13. J. P. Wolanczyk, Kenneth B. Storey, and John G. Baust, "Ice Nucleating Activity in the Blood of the Freeze-Tolerant Frog, *Rana sylvatica*," *Cryobiology* 27, no. 3 (June 1990), https://doi.org/10.1016/0011-2240(90)90032-y.

14. Muñoz et al., "Structure and Application of Antifreeze Proteins."

15. Tsui-Chin Huang, Jheng-Fong Lee, and Jyh-Yih Chen, "Pardaxin, an Antimicrobial Peptide, Triggers Caspase-Dependent and ROS-Mediated Apoptosis in HT-1080 cells," *Marine Drugs* 9, no. 10 (2011), https://doi.org/10.3390/md9101995.

BIRTH

1. Lin Römer and Thomas Scheibel, "The Elaborate Structure of Spider Silk: Structure and Function of a Natural High Performance Fiber," *Prion* 2, no. 4 (Oct.–Dec. 2008), https://doi.org/10.4161/pri.2.4.7490.

2. J. D. van Beek et al., "The Molecular Structure of Spider Dragline Silk: Folding and Orientation of the Protein Backbone," *Proceedings of the National Academy of Sciences USA* 99, no. 16 (Aug. 6, 2002), https://doi.org/10.1073/pnas.152162299.

3. Brandon Ho, Anastasia Baryshnikova, and Grant W. Brown, "Unification of Protein Abundance Datasets Yields a Quantitative *Saccharomyces cerevisiae* Proteome," *Cell Systems* 6, no. 2 (Feb. 28, 2018), https://doi.org/10.1016/j.cels.2017.12.004.

4. Klaus Weber and Mary Osborn, "Cytoskeleton: Definition, Structure and Gene Regulation," *Pathology Research and Practice* 175, no. 2–3 (1982), https://doi.org/10.1016/S0344-0338(82)80104-0.

5. Harald Herrmann and Ueli Aebi, "Intermediate Filaments: Structure and Assembly," *Cold Spring Harbor Perspectives in Biology* 8, no. 11 (Nov. 1, 2016), https://doi.org/10.1101/cshperspect.a018242.

6. Herrmann and Aebi, "Intermediate Filaments."

7. Wolfgang H. Goldmann, "Intermediate Filaments and Cellular Mechanics," *Cell Biology International* 42, no. 2 (Feb. 2018), https://doi.org/10.1002/cbin.10879.

8. Goldmann, "Intermediate Filaments and Cellular Mechanics."

9. Robert D. Goldman et al., "Intermediate Filaments: Versatile Building Blocks of Cell Structure," *Current Opinion in Cell Biology* 20, no. 1 (Feb. 2008), https://doi.org/10.1016/j.ceb.2007.11.003.

10. J. M. Gillespie and A. S. Inglis, "High-Sulphur Proteins as a Major Cause of Variation in Sulphur Content between Alpha-Keratins," *Nature* 207, no. 5003 (1965), https://doi.org/10.1016/0010-406X(65)90345-2.

11. John Squire, "Special Issue: The Actin-Myosin Interaction in Muscle: Background and Overview," *International Journal of Molecular Sciences* 20, no. 22 (Nov. 14, 2019), https://doi.org/10.3390/ijms20225715.

12. Emma J. van Bodegraven and Sandrine Etienne-Manneville, "Intermediate Filaments against Actomyosin: The David and Goliath of Cell Migration," *Current Opinion in Cell Biology* 66 (Oct. 2020), https://doi.org/10.1016/j.ceb.2020.05.006.

13. Squire, "Special Issue."

14. Van Bodegraven and Etienne-Manneville, "Intermediate Filaments against Actomyosin."

15. H. Lee Sweeney and Erika L. F. Holzbaur, "Motor Proteins," *Cold Spring Harbor Perspectives in Biology* 10, no. 5 (May 1, 2018), https://doi.org/10.1101/cshperspect.a021931.

16. Gary Borisy et al., "Microtubules: 50 Years on from the Discovery of Tubulin," *Nature Reviews Molecular Cell Biology* 17, no. 5 (Apr. 22, 2016), https://doi.org/10.1038/nrm .2016.45.

17. Holly V. Goodson and Erin M. Jonasson, "Microtubules and Microtubule-Associated Proteins," *Cold Spring Harbor Perspectives in Biology* 10, no. 6 (June 1, 2018), https://doi.org /10.1101/cshperspect.a022608.

18. Arshad Desai and Timothy J. Mitchison, "Microtubule Polymerization Dynamics," *Annual Review of Cell and Developmental Biology* 13 (1997), https://doi.org/10.1146/annurev.cellbio .13.1.83.

19. Megan K. DeBari et al., "Silk Fibroin as a Green Material," *ACS Biomaterials Science and Engineering* 7, no. 8 (Aug. 9, 2021), https://doi.org/10.1021/acsbiomaterials.1c00493.

20. Nick Lane, "The Unseen World: Reflections on Leeuwenhoek (1677) 'Concerning Little Animals,'" *Philosophical Transactions of the Royal Society B Biological Sciences* 370, no. 1666 (Apr. 19, 2015), https://doi.org/10.1098/rstb.2014.0344.

21. Shuichi Nakamura and Tohru Minamino, "Flagella-Driven Motility of Bacteria," *Biomolecules* 9, no. 7 (July 14, 2019), https://doi.org/10.3390/biom9070279.

22. Howard C. Berg, "The Rotary Motor of Bacterial Flagella," *Annual Review of Biochemistry* 72 (2003), https://doi.org/10.1146/annurev.biochem.72.121801.161737.

23. For more on the run-and-tumble movements of bacteria, see Chapter 3.

24. George H. Wadhams and Judith P. Armitage, "Making Sense of It All: Bacterial Chemotaxis," *Nature Reviews Molecular Cell Biology* 5, no. 12 (Dec. 2004), https://doi.org /10.1038/nrm1524.

25. Stuart Hameroff and Roger Penrose, "Consciousness in the Universe: A Review of the 'Orch OR' Theory," *Physics of Life Reviews* 11, no. 1 (Mar. 2014), https://doi.org/10.1016/j .plrev.2013.08.002.

AWAKENING

1. Yoshinori Shichida and Take Matsuyama, "Evolution of Opsins and Phototransduction," *Philosophical Transactions of the Royal Society B Biological Sciences* 364, no. 1531 (Oct. 12, 2009), https://doi.org/10.1098/rstb.2009.0051.

2. Babak Daneshfard, Behnam Dalfardi, and Golnoush S. Nezhad, "Ibn al-Haytham (965–1039 a.d.), the Original Portrayal of the Modern Theory of Vision," *Journal of Medical Biography* 24, no. 2 (May 2016), https://doi.org/10.1177/0967772014529050.

3. Shichida and Matsuyama, "Evolution of Opsins and Phototransduction."

4. Alexander L. Stubbs and Christopher W. Stubbs, "Spectral Discrimination in Color Blind Animals via Chromatic Aberration and Pupil Shape," *Proceedings of the National Academy of Sciences USA* 113, no. 29 (July 19, 2016), https://doi.org/10.1073/pnas.1524578113.

5. Peter G. Kevan, Lars Chittka, and Adrian G. Dyer, "Limits to the Salience of Ultraviolet: Lessons from Colour Vision in Bees and Birds," *Journal of Experimental Biology* 204, no. 14 (July 2001), https://doi.org/10.1242/jeb.204.14.2571.

6. Ilse M. Daly et al., "Dynamic Polarization Vision in Mantis Shrimps," *Nature Communications* 7 (July 12, 2016), https://doi.org/10.1038/ncomms12140.

7. Daly et al., "Dynamic Polarization Vision."

8. Atrouli Chatterjee et al., "An Introduction to Color-Changing Systems from the Cephalopod Protein Reflectin," *Bioinspiration and Biomimetics* 13, no. 4 (May 25, 2018), https://doi.org/10.1088/1748-3190/aab804.

9. Stubbs and Stubbs, "Spectral Discrimination."

10. Karl Edman et al., "High-Resolution X-Ray Structure of an Early Intermediate in the Bacteriorhodopsin Photocycle," *Nature* 401, no. 6755 (Oct. 21, 1999), https://doi.org/10.1038/44623.

11. Qin Wang and Chentao Lin, "Mechanisms of Cryptochrome-Mediated Photoresponses in Plants," *Annual Review of Plant Biology* 71 (Apr. 29, 2020), https://doi.org/10.1146/annurev-arplant-050718-100300.

12. Anthony R. Cashmore, "Cryptochromes: Enabling Plants and Animals to Determine Circadian Time," *Cell* 114, no. 5 (Sept. 5, 2003), https://www.ncbi.nlm.nih.gov/pubmed/13678578.

13. Cashmore, "Cryptochromes."

14. Alessandro Marchiori et al., "Coarse-Grained / Molecular Mechanics of the TAS2R38 Bitter Taste Receptor: Experimentally-Validated Detailed Structural Prediction of Agonist Binding," *PLoS One* 8, no. 5 (2013), https://doi.org/10.1371/journal.pone.0064675.

15. Chenyang Wu et al., "The Structure and Function of Olfactory Receptors," *Trends in Pharmacological Sciences* 45, no. 3 (Mar. 2024), https://doi.org/10.1016/j.tips.2024.01.004.

16. Marchiori et al., "Coarse-Grained / Molecular Mechanics."

17. Kamila Czarnecka et al., "Aspartame-True or False? Narrative Review of Safety Analysis of General Use in Products," *Nutrients* 13, no. 6 (June 7, 2021), https://doi.org/10.3390/nu13061957.

18. Yoav Gilad and Doron Lancet, "Population Differences in the Human Functional Olfactory Repertoire," *Molecular Biology and Evolution* 20, no. 3 (Mar. 2003), https://doi.org/10.1093/molbev/msg013.

19. Wu et al., "Structure and Function of Olfactory Receptors."

20. Jayant M. Pinto, "Olfaction," *Proceedings of the American Thoracic Society* 8, no. 1 (Mar. 2011), https://doi.org/10.1513/pats.201005-035RN.

21. C. Bushdid et al., "Humans Can Discriminate More Than 1 Trillion Olfactory Stimuli," *Science* 343, no. 6177 (Mar. 21, 2014), https://doi.org/10.1126/science.1249168.

22. Eileen K. Jenkins, Mallory T. DeChant, and Erin B. Perry, "When the Nose Doesn't Know: Canine Olfactory Function Associated with Health, Management, and Potential Links to Microbiota," *Frontiers in Veterinary Science* 5 (2018), https://doi.org/10.3389/fvets.2018.00056.

23. Shih-Feng Liu et al., "Sniffer Dogs Diagnose Lung Cancer by Recognition of Exhaled Gases: Using Breathing Target Samples to Train Dogs Has a Higher Diagnostic Rate Than Using Lung Cancer Tissue Samples or Urine Samples," *Cancers (Basel)* 15, no. 4 (Feb. 15, 2023), https://doi.org/10.3390/cancers15041234.

24. Claire L. Careaga et al., "Large Amplitude Twisting Motions of an Interdomain Hinge: A Disulfide Trapping Study of the Galactose-Glucose Binding Protein," *Biochemistry* 34, no. 9 (Mar. 7, 1995), https://doi.org/10.1021/bi00009a036.

25. A. J. Wolfe et al., "Reconstitution of Signaling in Bacterial Chemotaxis," *Journal of Bacteriology* 169, no. 5 (May 1, 1987), http://jb.asm.org/cgi/content/abstract/169/5/1878.

26. George H. Wadhams and Judith P. Armitage, "Making Sense of It All: Bacterial Chemotaxis," *Nature Reviews Molecular Cell Biology* 5, no. 12 (Dec. 2004), https://doi.org/10.1038/nrm1524.

27. Caroline Flegel et al., "Characterization of the Olfactory Receptors Expressed in Human Spermatozoa," *Frontiers in Molecular Biosciences* 2 (2015), https://doi.org/10.3389/fmolb.2015.00073.

28. Hiroshi Bandoh, Ikuhiro Kida, and Hiroshi Ueda, "Olfactory Responses to Natal Stream Water in Sockeye Salmon by BOLD fMRI," *PLoS One* 6, no. 1 (Jan. 17, 2011), https://doi.org/10.1371/journal.pone.0016051.

29. Beyza Ustun et al., "Flavor Sensing in Utero and Emerging Discriminative Behaviors in the Human Fetus," *Psychological Science* 33, no. 10 (Sept. 21, 2022), https://doi.org/10.1177/09567976221105460.

30. Julie A. Mennella, Coren P. Jagnow, and Gary K. Beauchamp, "Prenatal and Postnatal Flavor Learning by Human Infants," *Pediatrics* 107, no. 6 (Jun 2001), https://doi.org/10.1542/peds.107.6.e88.

31. Stéphanie A. Bayol, Samantha J. Farrington, and Neil C. Stickland, "A Maternal 'Junk Food' Diet in Pregnancy and Lactation Promotes an Exacerbated Taste for 'Junk Food' and a Greater Propensity for Obesity in Rat Offspring," *British Journal of Nutrition* 98, no. 4 (Oct. 2007), https://www.ncbi.nlm.nih.gov/pubmed/17697422.

32. Marchiori et al., "Coarse-Grained / Molecular Mechanics."

33. Daniel Hilger, Matthieu Masureel, and Brian K. Kobilka, "Structure and Dynamics of GPCR Signaling Complexes," *Nature Structural & Molecular Biology* 25, no. 1 (Jan. 2018), https://doi.org/10.1038/s41594-017-0011-7.

34. Hilger, Masureel, and Kobilka, "Structure and Dynamics of GPCR Signaling Complexes."

35. Sameer Jauhar, Philip J. Cowen, and Michael Browning, "Fifty Years On: Serotonin and Depression," *Journal of Psychopharmacology* 37, no. 3 (Mar. 2023), https://doi.org/10.1177/02698811231161813.

36. Markus Zweckstetter, Alexander Dityatev, and Evgeni Ponimaskin, "Structure of Serotonin Receptors: Molecular Underpinning of Receptor Activation and Modulation," *Signal Transduction and Targeted Therapy* 6, no. 1 (June 18, 2021), https://doi.org/10.1038/s41392-021-00668-3, https://www.ncbi.nlm.nih.gov/pubmed/34145221.

37. G. K. Aghajanian and G. J. Marek, "Serotonin and Hallucinogens," *Neuropsychopharmacology* 21, no. 2 supp. (Aug. 1999), https://doi.org/10.1016/S0893-133X(98)00135-3.

38. Catherine Maud et al., "The Role of Oxytocin Receptor Gene (OXTR) DNA Methylation (DNAm) in Human Social and Emotional Functioning: A Systematic Narrative Review," *BMC Psychiatry* 18, no. 1 (May 29, 2018), https://doi.org/10.1186/s12888-018-1740-9.

39. Alessandra de Felice, Simone Aureli, and Vittorio Limongelli, "Drug Repurposing on G Protein-Coupled Receptors Using a Computational Profiling Approach," *Frontiers in Molecular Biosciences* 8 (2021), https://doi.org/10.3389/fmolb.2021.673053.

40. Annie Handler and David D. Ginty, "The Mechanosensory Neurons of Touch and Their Mechanisms of Activation," *Nature Reviews Neuroscience* 22, no. 9 (Sept. 2021), https://doi .org/10.1038/s41583-021-00489-x.

41. Diana S. Stetson et al., "Effects of Age, Sex, and Anthropometric Factors on Nerve Conduction Measures," *Muscle & Nerve* 15, no. 10 (1992), https://doi.org/https://doi.org/10 .1002/mus.880151007.

42. Oxygen in its elemental form (O_2) can be attracted by a strong magnet when cooled and liquified, but in a protein, all the oxygen electrons are paired, rendering it non-magnetic.

43. Atticus Pinzon-Rodriguez, Staffan Bensch, and Rachel Muheim, "Expression Patterns of Cryptochrome Genes in Avian Retina Suggest Involvement of Cry4 in Light-Dependent Magnetoreception," *Journal of the Royal Society Interface* 15, no. 140 (Mar. 2018), https://doi.org/10.1098/rsif.2018.0058.

44. James A. Simmons, "Perception of Echo Phase Information in Bat Sonar," *Science* 204, no. 4399 (1979), https://www.science.org/doi/10.1126/science.451543; W. L. Au Whitlow, "Hearing in Whales and Dolphins: An Overview," in Whitlow W. L. Au, Arthur N. Popper, and Richard R. Fay, eds., *Hearing by Whales and Dolphins* (New York: Springer, 2000), https://link.springer.com/chapter/10.1007/978-1-4612-1150-1_1.

45. Nicholas W. Bellono, Duncan B. Leitch, and David Julius, "Molecular Tuning of Electroreception in Sharks and Skates," *Nature* 558, no. 7708 (2018), https://www.ncbi .nlm.nih.gov/pmc/articles/PMC6101975; Eric A. Newman and Peter H. Hartline, "The Infrared 'Vision' of Snakes," *Scientific American* 246, no. 3 (Mar. 1982).

46. William E. Cooper Jr. and Valentín Pérez-Mellado, "Location of Fruit Using Only Airborne Odor Cues by a Lizard," *Physiology & Behavior* 74, no. 3 (2001), https://podarcis.de/AF /Bibliografie/BIB_1801.pdf; Michael Garstang and Michael C. Kelley, "Understanding Animal Detection of Precursor Earthquake Sounds," *Animals* 7, no. 9 (2017), https://www .mdpi.com/2076-2615/7/9/66.

47. Ernst Mayr, *What Evolution Is* (New York: Basic Books, 2001).

TAKING FORM

1. David Eisenberg, "The Discovery of the Alpha-Helix and Beta-Sheet, the Principal Structural Features of Proteins," *Proceedings of the National Academy of Sciences USA* 100, no. 20 (Sept. 30, 2003), https://doi.org/10.1073/pnas.2034522100.

2. Samir S. Amr and Abdelghani Tbakhi, "Jabir ibn Hayyan," *Annals of Saudi Medicine* 27, no. 1 (Jan.–Feb. 2007), https://www.ncbi.nlm.nih.gov/pubmed/17337999.

3. Liz Kambas, "Antoine-Laurent Lavoisier's 'Sur la nature de l'eau': An Annotated English Translation," *Annals of Science* (Jan. 12, 2024), https://doi.org/10.1080/00033790.2023 .2289531.

4. "Periodic Table of Elements," American Chemistry Society website, https://www.acs.org /education/whatischemistry/periodictable.html.

5. Linus Pauling, *The Nature of the Chemical Bond and the Structure of Molecules and Crystals: An Introduction to Modern Structural Chemistry* (Ithaca, NY: Cornell University Press, 1939).

6. A. D. Hershey and Martha Chase, "Independent Functions of Viral Protein and Nucleic Acid in Growth of Bacteriophage," *Journal of General Physiology* 36, no. 1 (May 1952), https://doi.org/10.1085/jgp.36.1.39.

https://www.researchgate.net/publication/12321373_Folic_Acid_Nutritional
_Biochemistry_Molecular_Biology_and_Role_in_Disease_Processes.

28. Jing-Ke Weng and Clint Chapple, "The Origin and Evolution of Lignin Biosynthesis," *New Phytologist* 187, no. 2 (July 2010), https://doi.org/10.1111/j.1469-8137.2010.03327.x.

29. Grzegorz Janusz et al., "Lignin Degradation: Microorganisms, Enzymes Involved, Genomes Analysis and Evolution," *FEMS Microbiology Reviews* 41, no. 6 (Nov. 1, 2017), https://doi.org/10.1093/femsre/fux049.

30. Paolo Bombelli, Christopher J. Howe, and Federica Bertocchini, "Polyethylene Bio-Degradation by Caterpillars of the Wax Moth *Galleria mellonella*," *Current Biology* 27, no. 8 (Apr. 24, 2017), https://doi.org/10.1016/j.cub.2017.02.060.

TRANSFORMATION

1. Thomas Eisner et al., "Firefly 'Femmes Fatales' Acquire Defensive Steroids (Lucibufagins) from Their Firefly Prey," *Proceedings of the National Academy of Sciences of the USA* 94, no. 18 (Sept. 2, 1997), https://www.pnas.org/doi/full/10.1073/pnas.94.18.9723.

2. Thomas Eisner et al., "Lucibufagins: Defensive Steroids from the Fireflies *Photinus ignitus* and *P. marginellus* (Coleoptera: Lampyridae)," *Proceedings of the National Academy of Sciences USA* 75, no. 2 (Feb. 1978), https://doi.org/10.1073/pnas.75.2.905.

3. Brian C. Leavell et al., "Fireflies Thwart Bat Attack with Multisensory Warnings," *Science Advances* 4, no. 8 (Aug. 2018), https://doi.org/10.1126/sciadv.aat6601.

4. Eve Otjacques et al., "Bioluminescence in Cephalopods: Biodiversity, Biogeography and Research Trends," *Frontiers in Marine Science* 10 (June 27 2023), https://www.frontiersin.org/journals/marine-science/articles/10.3389/fmars.2023.1161049/full.

5. Anne K. Dunn, "*Vibrio fischeri* Metabolism: Symbiosis and Beyond," *Advances in Microbial Physiology* 61 (2012), https://doi.org/10.1016/B978-0-12-394423-8.00002-0, https://www.ncbi.nlm.nih.gov/pubmed/23046951.

6. Dunn, "*Vibrio fischeri* Metabolism."

7. Nick Lane, "The Unseen World: Reflections on Leeuwenhoek (1677) 'Concerning Little Animals,'" *Philosophical Transactions of the Royal Society B Biological Sciences* 370, no. 1666 (Apr. 19, 2015), https://doi.org/10.1098/rstb.2014.0344.

8. Jacques Poisson, "[Raphael Dubois, from Pharmacy to Bioluminescence]," *Revue d'histoire de la pharmacie* (Paris) 58, no. 365 (Apr. 2010), https://www.ncbi.nlm.nih.gov/pubmed/20533808.

9. Matthew P. Davis, John S. Sparks, and W. Leo Smith, "Repeated and Widespread Evolution of Bioluminescence in Marine Fishes," *PLoS One* 11, no. 6 (2016), https://doi.org/10.1371/journal.pone.0155154.

10. Jérôme Delroisse et al., "Leaving the Dark Side? Insights into the Evolution of Luciferases," *Frontiers in Marine Science* 8 (June 30, 2021), https://www.frontiersin.org/journals/marine-science/articles/10.3389/fmars.2021.673620/full.

11. Timothy R. Fallon et al., "Firefly Genomes Illuminate Parallel Origins of Bioluminescence in Beetles," *eLife* 7 (Oct. 16, 2018), https://doi.org/10.7554/eLife.36495.

12. Fallon et al., "Firefly Genomes."

13. Davis, Sparks, and Smith, "Repeated and Widespread Evolution."

14. Davis, Sparks, and Smith, "Repeated and Widespread Evolution."

15. Delroisse et al., "Leaving the Dark Side?"

16. Delroisse et al., "Leaving the Dark Side?"

17. V. R. Bevilaqua et al., "*Phrixotrix* Luciferase and 6'-Aminoluciferins Reveal a Larger Luciferin Phenolate Binding Site and Provide Novel Far-Red Combinations for Bioimaging Purposes," *Scientific Reports* 9 (June 21, 2019), https://www.nature.com/articles/s41598-019-44534-3.

18. Aleksandra S. Tsarkova, "Luciferins under Construction: A Review of Known Biosynthetic Pathways," *Frontiers in Ecology and Evolution* 9 (Sept. 20, 2021), https://www.frontiersin .org/journals/ecology-and-evolution/articles/10.3389/fevo.2021.667829/full.

19. Eloi P. Coutant et al., "Gram-Scale Synthesis of Luciferins Derived from Coelenterazine and Original Insights into Their Bioluminescence Properties," *Organic and Biomolecular Chemistry* 17, no. 15 (Apr. 10, 2019), https://doi.org/10.1039/c9ob00459a.

20. Tsarkova, "Luciferins under Construction."

21. V. R. Viviani, "The Origin, Diversity, and Structure Function Relationships of Insect Luciferases," *Cellular and Molecular Life Sciences* 59, no. 11 (Nov. 2002), https://www .researchgate.net/publication/10947465_The_origin_diversity_and_structure_function _relationships_of_insect_luciferases.

22. Jeffrey L. Bose et al., "Bioluminescence in *Vibrio fischeri* Is Controlled by the Redox-Responsive Regulator ArcA," *Molecular Microbiology* 65, no. 2 (July 2007), https://doi.org /10.1111/j.1365-2958.2007.05809.x.

23. M. B. Miller and B. L. Bassler, "Quorum Sensing in Bacteria," *Annual Review of Microbiology* 55 (2001), https://pubmed.ncbi.nlm.nih.gov/11544353.

24. David P. Clark, Nanette Jean Pazdernik, and Michelle R. McGehee, *Molecular Biology*, 3rd ed. (London: Academic Press, 2019).

25. G. S. Timmins et al., "Firefly Flashing Is Controlled by Gating Oxygen to Light-Emitting Cells," *Journal of Experimental Biology* 204, no. 16 (Aug. 2001), https://journals.biologists .com/jeb/article/204/16/2795/32847/firefly-flashing-is-controlled-by-gating-oxygen-to.

26. Stephanie A. Smith, Richard J. Travers, and James H. Morrissey, "How It All Starts: Initiation of the Clotting Cascade," *Critical Reviews in Biochemistry and Molecular Biology* 50, no. 4 (2015), https://doi.org/10.3109/10409238.2015.1050550.

27. M. Ormö et al., "Crystal Structure of the *Aequorea victoria* Green Fluorescent Protein," *Science* 273, no. 5280 (Sept. 6, 1996), https://pubmed.ncbi.nlm.nih.gov/8703075.

28. Valérie Baubet et al., "Chimeric Green Fluorescent Protein-Aequorin as Bioluminescent Ca2+ Reporters at the Single-Cell Level," *Proceedings of the National Academy of Sciences USA* 97, no. 13 (June 20, 2000), https://doi.org/10.1073/pnas.97.13.7260.

29. Ormö et al., "Crystal Structure of the *Aequorea victoria*."

30. Marc Zimmer, "GFP: From Jellyfish to the Nobel Prize and Beyond," *Chemical Society Reviews* 38, no. 10 (2009), https://doi.org/10.1039/b904023d.

31. O. Shimomura, "Discovery of Green Fluorescent Protein (GFP) (Nobel Lecture)," *Angewandte Chemie-International Edition* 48, no. 31 (2009), https://doi.org/10.1002/anie .200902240.

32. Zimmer, "GFP."

33. N. M. Rusan, C. J. Fagerstrom, A. M. Yvon, and P. Wadsworth, "Cell Cycle–Dependent Changes in Microtubule Dynamics in Living Cells Expressing Green Fluorescent Protein-Alpha Tubulin," *Molecular Biology of the Cell* 12, no. 4 (Apr. 2001): 971–980, https://pubmed.ncbi.nlm.nih.gov/11294900/.

34. Anny Follenius-Wund et al., "Fluorescent Derivatives of the GFP Chromophore Give a New Insight into the GFP Fluorescence Process," *Biophysical Journal* 85, no. 3 (Sept. 2003), https://doi.org/10.1016/S0006-3495(03)74612-8.

35. Tamily A. Weissman and Y. Albert Pan, "Brainbow: New Resources and Emerging Biological Applications for Multicolor Genetic Labeling and Analysis," *Genetics* 199, no. 2 (Feb. 2015), https://doi.org/10.1534/genetics.114.172510.

36. Ruxana T. Sadikot and Timothy S. Blackwell, "Bioluminescence Imaging," *Proceedings of the American Thoracic Society* 2, no. 6 (2005), https://doi.org/10.1513/pats.200507-067DS.

37. Serikbai K. Abilev et al., "Bacterial Lux Biosensors in Genotoxicological Studies," *Biosensors* 13, no. 5 (Apr. 29, 2023), https://doi.org/10.3390/bios13050511.

REMEMBERING

1. Dan Torre, *Carnivorous Plants* (London: Reaktion Books, 2023).

2. Charles Darwin and Francis Darwin, *Insectivorous Plants* (London: Murray, 1875).

3. Rainer Hedrich and Erwin Neher, "Venus Flytrap: How an Excitable, Carnivorous Plant Works," *Trends in Plant Science* 23, no. 3 (2018), https://www.researchgate.net/publication/322411142_Venus_Flytrap_How_an_Excitable_Carnivorous_Plant_Works.

4. Hiraku Suda et al., "Calcium Dynamics during Trap Closure Visualized in Transgenic Venus Flytrap," *Nature Plants* 6, no. 10 (Oct. 2020), https://doi.org/10.1038/s41477-020-00773-1.

5. Rob Crane, "The Most Addictive Drug, the Most Deadly Substance: Smoking Cessation Tactics for the Busy Clinician," *Primary Care: Clinics in Office Practice* 34, no. 1 (2007); Stephen S. Hecht, "Tobacco Carcinogens, Their Biomarkers and Tobacco-Induced Cancer," *Nature Reviews Cancer* 3, no. 10 (2003), https://www.researchgate.net/publication/231585301_Tobacco_carcinogens_their_biomarkers_and_tobacco-induced_cancer.

6. Gerd P. Pfeifer, Young-Hyun You, and Ahmad Besaratinia, "Mutations Induced by Ultraviolet Light," *Mutation Research / Fundamental and Molecular Mechanisms of Mutagenesis* 571, no. 1–2 (2005), https://www.sciencedirect.com/science/article/abs/pii/S0027510704004804.

7. Iñigo Martincorena and Peter J. Campbell, "Somatic Mutation in Cancer and Normal Cells," *Science* 349, no. 6255 (2015), https://www.researchgate.net/publication/283028168_Somatic_mutation_in_cancer_and_normal_cells.

8. Thomas A. Kunkel, "DNA Replication Fidelity," *Journal of Biological Chemistry* 279, no. 17 (2004), https://www.jbc.org/article/S0021-9258(19)75503-3/fulltext.

9. Beth Gibson et al., "The Distribution of Bacterial Doubling Times in the Wild," *Proceedings of the Royal Society B* 285, no. 1880 (2018), https://royalsocietypublishing.org/doi/10.1098/rspb.2018.0789.

10. Ravi R. Iyer et al., "DNA Mismatch Repair: Functions and Mechanisms," *Chemical Reviews* 106, no. 2 (2006), https://pubmed.ncbi.nlm.nih.gov/16464007.

11. Guo-Min Li, "A Personal Tribute to 2015 Nobel Laureate Paul Modrich," *DNA Repair* 37 (2016), https://pubmed.ncbi.nlm.nih.gov/26861181.

12. Gregory Beck and Gail S. Habicht, "Immunity and the Invertebrates," *Scientific American* 275, no. 5 (1996): 60–66, http://www.jstor.org/stable/24993447.

13. Shane Crotty, "A Brief History of T Cell Help to B Cells," *Nature Reviews Immunology* 15, no. 3 (2015), https://www.ncbi.nlm.nih.gov/pmc/articles/PMC4414089.

14. Nigel Chaffey, "Alberts, B., Johnson, A., Lewis, J., Raff, M., Roberts, K. and Walter, P. Molecular Biology of the Cell, 4th ed.," *Annals of Botany* 91, no. 3 (Feb. 2003), https://www.ncbi.nlm.nih.gov/pmc/articles/PMC4244961.

15. Lauren Thau, Edinen Asuka, and Kunal Mahajan, "Physiology, Opsonization," *StatPearls* (2023), https://www.ncbi.nlm.nih.gov/books/NBK534215.

16. Tomohiro Kurosaki, Kohei Kometani, and Wataru Ise, "Memory B Cells," *Nature Reviews Immunology* 15, no. 3 (2015), https://www.nature.com/articles/nri3802.

17. P. D. Ellner, "Smallpox: Gone But Not Forgotten," *Infection* 26, no. 5 (1998), https://colab.ws/articles/10.1007%2Fbf02962244.

18. Geoff Brumfiel, "A 300-Year-Old Tale of One Woman's Quest to Stop a Deadly Virus," podcast audio, All Things Considered, National Public Radio, Mar. 22, 2021, https://www.npr.org/sections/health-shots/2021/03/08/972978143/a-300-year-old-tale-of-one-womans-quest-to-stop-a-deadly-virus.

19. Donald K. Milton, "What Was the Primary Mode of Smallpox Transmission? Implications for Biodefense," *Frontiers in Cellular and Infection Microbiology* 2 (2012), https://pubmed.ncbi.nlm.nih.gov/23226686.

20. Jo Willett, *The Pioneering Life of Mary Wortley Montagu: Scientist and Feminist* (South Yorkshire, UK: Pen and Sword History, 2021).

21. Erin Blakemore, "How an Enslaved African Man in Boston Helped Save Generations from Smallpox," *History.com,* Apr. 8, 2021 (updated), https://www.history.com/news/smallpox-vaccine-onesimus-slave-cotton-mather.

22. Arthur Boylston, "The Origins of Inoculation," *Journal of the Royal Society of Medicine* 105, no. 7 (July 2012), https://doi.org/10.1258/jrsm.2012.12k044.

23. Blakemore, "How an Enslaved African Man."

24. Kendall A. Smith, "Edward Jenner and the Small Pox Vaccine," *Frontiers in Immunology* 2, no. 21 (2011), https://www.ncbi.nlm.nih.gov/pmc/articles/PMC3342363.

25. Jason Beaubien, "A Cow Head Will Not Erupt from Your Body if You Get a Smallpox Vaccine," Goats and Soda, National Public Radio, Jan. 7, 2015, https://www.npr.org/sections/goatsandsoda/2015/01/07/375598652/a-cow-head-will-not-erupt-from-your-body-if-you-get-a-smallpox-vaccine.

26. Donald A. Henderson, "Principles and Lessons from the Smallpox Eradication Programme," *Bulletin of the World Health Organization* 65, no. 4 (1987).

27. Kamran Badizadegan, Dominika A. Kalkowska, and Kimberly M. Thompson, "Polio by the Numbers—A Global Perspective," *Journal of Infectious Diseases* 226, no. 8 (2022), https://pubmed.ncbi.nlm.nih.gov/35415741.

28. Denise L Doolan, Carlota Dobaño, and J. Kevin Baird, "Acquired Immunity to Malaria," *Clinical Microbiology Reviews* 22, no. 1 (2009), https://pubmed.ncbi.nlm.nih.gov/19136431.

29. Wenhan Shao et al., "Evolution of Influenza A Virus by Mutation and Re-Assortment," *International Journal of Molecular Sciences* 18, no. 8 (2017), https://www.ncbi.nlm.nih.gov/pmc/articles/PMC5578040.

30. Matthew K. Waldor, "Bacteriophage Biology and Bacterial Virulence," *Trends in Microbiology* 6, no. 8 (1998), https://www.cell.com/trends/microbiology/abstract/S0966-842X(98)01320-1.

31. Philippe Horvath and Rodolphe Barrangou, "CRISPR / Cas, the Immune System of Bacteria and Archaea," *Science* 327, no. 5962 (Jan. 8, 2010), https://doi.org/10.1126/science.1179555.

32. Josiane E. Garneau et al., "The CRISPR / Cas Bacterial Immune System Cleaves Bacteriophage and Plasmid DNA," *Nature* 468, no. 7320 (2010), https://doi.org/10.1038/nature09523.

33. Horvath and Barrangou, "CRISPR / Cas."

34. Tzu-Chieh Ho et al., "Scaffold-Mediated CRISPR-Cas9 Delivery System for Acute Myeloid Leukemia Therapy," *Science Advances* 7, no. 21 (2021), https://www.science.org/doi/10.1126

/sciadv.abg3217. See also US Food & Drug Administration, "FDA Approves First Gene Therapies to Treat Patients with Sickle Cell Disease," press release, Dec. 8, 2023, https://www.fda.gov/news-events/press-announcements/fda-approves-first-gene-therapies -treat-patients-sickle-cell-disease.

35. Boylston, "Origins of Inoculation."

DEFIANCE

1. Wesam M. Salama and Khadiga M. Sharshar, "Surveillance Study on Scorpion Species in Egypt and Comparison of Their Crude Venom Protein Profiles," *Journal of Basic & Applied Zoology* 66, no. 2 (2013) https://www.sciencedirect.com/science/article/pii/ S2090989613000416.

2. Vivian Lee and Nada Rashwan, "Plagues Strike Egypt: Sudden Floods, Then 4-Inch Scorpions Called Deathstalkers," *New York Times*, Nov. 15, 2021, https://www.nytimes.com /2021/11/15/world/middleeast/scorpions-egypt.html.

3. P. G. Ojeda, C. K. Wang, and D. J. Craik, "Chlorotoxin: Structure, Activity, and Potential Uses in Cancer Therapy," *Peptide Science* 106, no. 1 (Jan. 2016), https://doi.org/10.1002/bip .22748.

4. Nicholas Roberts et al., "Global Mortality of Snakebite Envenoming between 1990 and 2019," *Nature Communications* 13, no. 1 (2022), https://pubmed.ncbi.nlm.nih.gov /36284094.

5. World Health Organization, "Zoonotic Disease Control: Baseline Epidemiological Study on Snake-Bite Treatment and Management," *Weekly Epidemiological Record* 62, no. 42 (1987), https://iris.who.int/handle/10665/226464.

6. Christian Betzel et al., "The Refined Crystal Structure of Alpha-Cobratoxin from Naja Naja Siamensis at 2.4-A Resolution," *Journal of Biological Chemistry* 266, no. 32 (1991), https://www.researchgate.net/publication/348253390_The_refined_crystal_structure_of _alpha-cobratoxin_from_Naja_naja_siamensis_at_24-A_resolution.

7. S. J. Burden, H. C. Hartzell, and D. Yoshikami, "Acetylcholine Receptors at Neuromuscular Synapses—Phylogenetic Differences Detected by Snake Alpha-Neurotoxins," *Proceedings of the National Academy of Sciences USA* 72, no. 8 (1975), https://www.pnas.org/doi/10.1073 /pnas.72.8.3245.

8. Andreas H. Laustsen et al., "Unveiling the Nature of Black Mamba (*Dendroaspis Polylepis*) Venom through Venomics and Antivenom Immunoprofiling: Identification of Key Toxin Targets for Antivenom Development," *Journal of Proteomics* 119 (Apr. 24, 2015), https://www .sciencedirect.com/science/article/abs/pii/S1874391915000561?via%3Dihub.

9. Eiko Ueno and Philip Rosenberg, "Mechanism of Action of β-bungarotoxin, a Presynaptically Acting Phospholipase A2 Neurotoxin: Its Effect on Protein Phosphorylation in Rat Brain Synaptosomes," *Toxicon* 34, no. 11–12 (1996), https://www.sciencedirect.com/science /article/abs/pii/S0041010196001134.

10. Karen L. Bell, Struan K. Sutherland, and Wayne C. Hodgson, "Some Pharmacological Studies of Venom from the Inland Taipan (*Oxyuranus microlepidotus*)," *Toxicon* 36, no. 1 (1998), https://pubmed.ncbi.nlm.nih.gov/9604283.

11. Carol Clarke et al., "Oxylepitoxin-1, a Reversible Neurotoxin from the Venom of the Inland Taipan (*Oxyuranus microlepidotus*)," *Peptides* 27, no. 11 (2006), https://www .sciencedirect.com/science/article/abs/pii/S0196978106002919.

12. Bryan G. Fry, "From Genome to 'Venome': Molecular Origin and Evolution of the Snake Venom Proteome Inferred from Phylogenetic Analysis of Toxin Sequences and Related Body Proteins," *Genome Research* 15, no. 3 (Mar. 2005), https://doi.org/10.1101/gr.3228405.

13. Lynne A. Isbell, "Snakes as Agents of Evolutionary Change in Primate Brains," *Journal of Human Evolution* 51, no. 1 (July 2006), https://doi.org/10.1016/j.jhevol.2005.12.012.

14. Bryan G. Fry et al., "A Central Role for Venom in Predation by *Varanus komodoensis* (Komodo Dragon) and the Extinct Giant *Varanus* (Megalania) *priscus*," *Proceedings of the National Academy of Sciences USA* 106, no. 22 (June 2, 2009), https://doi.org/10.1073/pnas.0810883106.

15. Emine Kocyigit et al., "Plant Toxic Proteins: Their Biological Activities, Mechanism of Action and Removal Strategies," *Toxins* 15, no. 6 (2023), https://www.researchgate.net /publication/371006193_Plant_Toxic_Proteins_Their_Biological_Activities_Mechanism _of_Action_and_Removal_Strategies.

16. J. Michael Lord, Lynne M. Roberts, and Jon D. Robertus, "Ricin: Structure, Mode of Action, and Some Current Applications," *FASEB Journal* 8, no. 2 (Feb. 1994), https://doi .org/DOI 10.1096 / fasebj.8.2.8119491.

17. Natalia Sowa-Rogozińska et al., "Intracellular Transport and Cytotoxicity of the Protein Toxin Ricin," *Toxins* 11, no. 6 (2019), https://pubmed.ncbi.nlm.nih.gov/31216687.

18. Alan J. Kohn, "Piscivorous Gastropods of the Genus *Conus*," *Proceedings of the National Academy of Sciences* 42, no. 3 (1956), https://www.ncbi.nlm.nih.gov/pmc/articles/PMC528241.

19. B. M. Olivera, "A Serendipitous Path to Pharmacology," *Annual Review of Pharmacology and Toxicology* 61 (2021), https://doi.org/10.1146/annurev-pharmtox-030320-113510.

20. S. W. A. Himaya and Richard J. Lewis, "Venomics-Accelerated Cone Snail Venom Peptide Discovery," *International Journal of Molecular Sciences* 19, no. 3 (2018), https://www.ncbi .nlm.nih.gov/pmc/articles/PMC5877649.

21. Joshua P. Torres et al., "Small-Molecule Mimicry Hunting Strategy in the Imperial Cone Snail, *Conus imperialis*," *Science Advances* 7, no. 11 (2021), https://www.science.org/doi/10 .1126/sciadv.abf2704.

22. John G. Menting et al., "A Minimized Human Insulin-Receptor-Binding Motif Revealed in a *Conus geographus* Venom Insulin," *Nature Structural & Molecular Biology* 23, no. 10 (Oct. 2016), https://doi.org/10.1038/nsmb.3292.

23. Joseph G McGivern, "Ziconotide: A Review of Its Pharmacology and Use in the Treatment of Pain," *Neuropsychiatric Disease and Treatment* 3, no. 1 (2007), https://www.ncbi.nlm.nih .gov/pmc/articles/PMC2654521.

24. Noemi Sanchez-Campos, Johanna Bernaldez-Sarabia, and Alexei F. Licea-Navarro, "Conotoxin Patenting Trends in Academia and Industry," *Marine Drugs* 20, no. 8 (2022), https://www.mdpi.com/1660-3397/20/8/531.

25. Dongrui Wang et al., "Chlorotoxin-Directed CAR T Cells for Specific and Effective Targeting of Glioblastoma," *Science Translational Medicine* 12, no. 533 (2020), https:// pubmed.ncbi.nlm.nih.gov/32132216.

26. Cho Yeow Koh and R. Manjunatha Kini, "From Snake Venom Toxins to Therapeutics— Cardiovascular Examples," *Toxicon* 59, no. 4 (Mar. 15, 2012), https://doi.org/10.1016/j .toxicon.2011.03.017.

27. Philip Lazarovici, Cezary Marcinkiewicz, and Peter I. Lelkes, "From Snake Venom's Disintegrins and C-Type Lectins to Anti-Platelet Drugs," *Toxins* 11, no. 5 (2019), https://www.mdpi.com/2072-6651/11/5/303.

28. Bryan J. Berube and Juliane B. Wardenburg, "*Staphylococcus α*-Toxin: Nearly a Century of Intrigue," *Toxins* 5, no. 6 (June 2013), https://doi.org/10.3390/toxins5061140.

29. Marta Michalska and Philipp Wolf, "*Pseudomonas* Exotoxin A: Optimized by Evolution for Effective Killing," *Frontiers in Microbiology* 6 (2015), https://www.ncbi.nlm.nih.gov/pmc /articles/PMC4584936.

30. Bernard Poulain and Michel R Popoff, "Why Are Botulinum Neurotoxin-Producing Bacteria so Diverse and Botulinum Neurotoxins so Toxic?," *Toxins* 11, no. 1 (2019), https://www.ncbi.nlm.nih.gov/pmc/articles/PMC6357194.

31. B. Zane Horowitz, "Botulinum Toxin," *Critical Care Clinics* 21, no. 4 (2005), https://ohsu .elsevierpure.com/en/publications/botulinum-toxin-2.

32. Iqbal Multani et al., "Botulinum Toxin in the Management of Children with Cerebral Palsy," *Pediatric Drugs* 21, no. 4 (2019), https://pubmed.ncbi.nlm.nih.gov/31257556.

33. Werner J. Becker, "Botulinum Toxin in the Treatment of Headache," *Toxins* 12, no. 12 (2020), https://www.ncbi.nlm.nih.gov/pmc/articles/PMC7766412.

34. Bagus Komang Satriyasa, "Botulinum toxin (Botox) A for Reducing the Appearance of Facial Wrinkles: A Literature Review of Clinical Use and Pharmacological Aspect," *Clinical, Cosmetic and Investigational Dermatology* (2019), https://www.ncbi.nlm.nih.gov /pmc/articles/PMC6489637.

35. Poulain and Popoff, "Why Are Botulinum Neurotoxin-Producing Bacteria so Diverse?"

36. J. Acosta et al., "Cloning and Functional Characterization of Three Novel Antimicrobial Peptides from Tilapia (*Oreochromis niloticus*)," *Fish & Shellfish Immunology* 34, no. 6 (June 2013), https://www.sciencedirect.com/science/article/pii/S1050464813002714 ?via%3Dihub.

37. J. K. Boparai and P. K. Sharma, "Mini Review on Antimicrobial Peptides, Sources, Mechanism and Recent Applications," *Protein and Peptide Letters* 27, no. 1 (2020), https://www.ncbi.nlm.nih.gov/pmc/articles/PMC6978648.

38. Jiaqi Xuan et al., "Antimicrobial Peptides for Combating Drug-Resistant Bacterial Infections," *Drug Resistance Updates* 68 (2023), https://pubmed.ncbi.nlm.nih.gov /36905712.

39. J. J. Schneider et al., "Human Defensins," *Journal of Molecular Medicine* 83, no. 8 (Aug 2005), https://doi.org/10.1007/s00109-005-0657-1.

DYING

1. Paolo d'Errico and Melanie Meyer-Luehmann, "Mechanisms of Pathogenic Tau and Abeta Protein Spreading in Alzheimer's Disease," *Frontiers in Aging Neuroscience* 12 (2020), https://doi.org/10.3389/fnagi.2020.00265.

2. Xiaojuan Sun, Wei-Dong Chen, and Yan-Dong Wang, "Beta-Amyloid: The Key Peptide in the Pathogenesis of Alzheimer's Disease," *Frontiers in Pharmacology* 6 (2015), https://doi .org/10.3389/fphar.2015.00221.

3. Harald Hampel et al., "The Amyloid-beta Pathway in Alzheimer's Disease," *Molecular Psychiatry* 26, no. 10 (Oct 2021), https://doi.org/10.1038/s41380-021-01249-0, https://www .ncbi.nlm.nih.gov/pubmed/34456336.

4. Fong Ping Chong et al., "Tau Proteins and Tauopathies in Alzheimer's Disease," *Cellular and Molecular Neurobiology* 38, no. 5 (Jul 2018), https://doi.org/10.1007/s10571-017-0574-1.

5. Simran J. Kaur, Stephanie R. McKeown, and Shazia Rashid, "Mutant SOD1 Mediated Pathogenesis of Amyotrophic Lateral Sclerosis," *Gene* 577, no. 2 (Feb. 15, 2016), https://doi.org/10.1016/j.gene.2015.11.049.

6. Lucia Banci et al., "SOD1 and Amyotrophic Lateral Sclerosis: Mutations and Oligomerization," *PLoS One* 3, no. 2 (Feb. 27, 2008), https://doi.org/10.1371/journal.pone.0001677.

7. Michael P. Alpers, "Review: The Epidemiology of Kuru: Monitoring the Epidemic from Its Peak to Its End," *Philosophical Transactions of the Royal Society B Biological Sciences* 363, no. 1510 (Nov. 27, 2008), https://doi.org/10.1098/rstb.2008.0071.

8. J. T. Whitfield et al., "Cultural Factors That Affected the Spatial and Temporal Epidemiology of Kuru," *Royal Society Open Science* 4, no. 1 (Jan 2017), https://www.ncbi.nlm.nih.gov/pmc/articles/PMC5319347.

9. Beata Sikorska and Pawel P. Liberski, "Human Prion Diseases: From Kuru to Variant Creutzfeldt-Jakob Disease," *Subcellular Biochemistry* 65 (2012), https://doi.org/10.1007/978-94-007-5416-4_17.

10. David C. Bolton, Michael P. McKinley, and Stanley B. Prusiner, "Identification of a Protein That Purifies with the Scrapie Prion," *Science* 218, no. 4579 (Dec. 24, 1982), https://doi.org/10.1126/science.6815801.

11. Nathan J. Cobb and Witold K. Surewicz, "Prion Diseases and Their Biochemical Mechanisms," *Biochemistry* 48, no. 12 (Mar. 31, 2009), https://doi.org/10.1021/bi900108v.

12. Ian Sample, "Should We Still Be Worried?," *The Guardian*, Jan. 10, 2007, https://www.theguardian.com/society/2007/jan/10/health.bse.

13. Sikorska and Liberski, "Human Prion Diseases."

14. Ainslie Tisdale et al., "The IDeaS initiative: Pilot Study to Assess the Impact of Rare Diseases on Patients and Healthcare Systems," *Orphanet Journal of Rare Diseases* 16, no. 1 (Oct. 22, 2021), https://doi.org/10.1186/s13023-021-02061-3.

15. Hartmut Grasemann and Felix Ratjen, "Cystic Fibrosis," *New England Journal of Medicine* 389, no. 18 (Nov. 2, 2023), https://doi.org/10.1056/NEJMra2216474.

16. Anne Vankeerberghen, Harry Cuppens, and Jean-Jacques Cassiman, "The Cystic Fibrosis Transmembrane Conductance Regulator: An Intriguing Protein with Pleiotropic Functions," *Journal of Cystic Fibrosis* 1, no. 1 (Mar. 2002), https://doi.org/10.1016/s1569-1993(01)00003-0.

17. Fernando A. L. Marson, Carmen S. Bertuzzo, and José D. Ribeiro, "Classification of *CFTR* Mutation Classes," *Lancet Respiratory Medicine* 4, no. 8 (Aug. 2016), https://doi.org/10.1016/S2213-2600(16)30188-6.

18. Montserrat Arrasate and Steven Finkbeiner, "Protein Aggregates in Huntington's Disease," *Experimental Neurology* 238, no. 1 (Nov. 2012), https://doi.org/10.1016/j.expneurol.2011.12.013.

19. Danny M. Hatters, "Protein Misfolding Inside Cells: The Case of Huntingtin and Huntington's Disease," *IUBMB Life* 60, no. 11 (Nov. 2008), https://doi.org/10.1002/iub.111.

20. Peg C. Nopoulos, "Special Issue: Juvenile Onset Huntington's Disease," *Brain Sciences* 10, no. 9 (Sept. 20, 2020), https://doi.org/10.3390/brainsci10090652.

21. Arrasate and Finkbeiner, "Protein Aggregates in Huntington's Disease."

22. M. A. South, "David the Bubble Boy: Some Lessons He Has Taught Us," *Cutis* 19, no. 5 (May 1977), https://www.ncbi.nlm.nih.gov/pubmed/301080.

23. Ivan K. Chinn and William T. Shearer, "Severe Combined Immunodeficiency Disorders," *Immunology and Allergy Clinics of North America* 35, no. 4 (Nov. 2015), https://doi.org/10.1016/j.iac.2015.07.002.

24. Michael Hershfield and Teresa Tarrant, "Adenosine Deaminase Deficiency," in M. P. Adam et al., eds., *GeneReviews* (Seattle: University of Washington, 1993).

25. Ramez Naam, *More Than Human* (New York: Broadway Books, 2005).

26. P. Bundock and P. J. Hooykaas, "Integration of *Agrobacterium tumefaciens*T-DNA in the *Saccharomyces cerevisiae* genome by Illegitimate Recombination," *Proceedings of the National Academy of Sciences*, 93, no. 26 (1996): 15272–15275. https://doi.org/10.1073/pnas.93.26.15272.

27. Fathema Uddin, Charles M. Rudin, and Triparna Sen, "CRISPR Gene Therapy: Applications, Limitations, and Implications for the Future," *Frontiers in Oncology* 10 (Aug. 7, 2020), https://www.ncbi.nlm.nih.gov/pmc/articles/PMC7427626.

28. Cormac Sheridan, "The World's First CRISPR Therapy Is Approved: Who Will Receive It?," *Nature Biotechnology* 42, no. 1 (Jan 2024), https://doi.org/10.1038/d41587-023-00016-6.

29. Henry T. Greely, "CRISPR'd Babies: Human Germline Genome Editing in the 'He Jiankui Affair'," *Journal of Law and Biosciences* 6, no. 1 (Oct. 2019), https://doi.org/10.1093/jlb/lsz010.

30. Sheldon Krimsky, "Ten Ways in Which He Jiankui Violated Ethics," *Nature Biotechnology* 37, no. 1 (Jan. 3, 2019), https://doi.org/10.1038/nbt.4337.

RESURRECTION

1. Diane B. Paul, "Darwin, Social Darwinism and Eugenics," *Cambridge Companion to Darwin* online,2006,https://www.dianebpaul.com/uploads/2/3/2/9/23295024/darwin_social_darwinism_and_eugenics.pdf.

2. Franklin M. Harold, *In Search of Cell History: The Evolution of Life's Building Blocks* (Chicago: University of Chicago Press, 2019).

3. Mark Farrugia and Byron Baron, "The Role of TNF-Alpha in Rheumatoid Arthritis: A Focus on Regulatory T Cells," *Journal of Clinical and Translational Research* 2, no. 3 (Nov. 10, 2016), https://www.ncbi.nlm.nih.gov/pubmed/30873466.

4. Daniel E. Furst et al., "Adalimumab, a Fully Human Anti Tumor Necrosis Factor–Alpha Monoclonal Antibody, and Concomitant Standard Antirheumatic Therapy for the Treatment of Rheumatoid Arthritis: Results of STAR (Safety Trial of Adalimumab in Rheumatoid Arthritis)," *Journal of Rheumatology* 30, no. 12 (Dec. 2003), https://www.ncbi.nlm.nih.gov/pubmed/14719195.

5. Ron Sender et al., "The Total Mass, Number, and Distribution of Immune Cells in the Human Body," *Proceedings of the National Academy of Sciences* 120, no. 44 (2023), https://www.pnas.org/doi/full/10.1073/pnas.2308511120.

6. Tineke Cantaert et al., "Activation-Induced Cytidine Deaminase Expression in Human B Cell Precursors Is Essential for Central B Cell Tolerance," *Immunity* 43, no. 5 (2015), https://www.sciencedirect.com/science/article/pii/S1074761315004021.

7. Charles Janeway et al., *Immunobiology: The Immune System in Health and Disease*, vol. 2 (New York: Garland, 2001).

8. Windy Dean-Colomb and Francisco J. Esteva, "Her2-Positive Breast Cancer: Herceptin and Beyond," *European Journal of Cancer* 44, no. 18 (Dec. 2008), https://doi.org/10.1016/j .ejca.2008.09.013, https://www.ncbi.nlm.nih.gov/pubmed/19022660.

9. Edith A. Perez et al., "Trastuzumab plus Adjuvant Chemotherapy for Human Epidermal Growth Factor Receptor 2–Positive Breast Cancer: Planned Joint Analysis of Overall Survival from NSABP B-31 and NCCTG N9831," *Journal of Clinical Oncology* 32, no. 33 (2014), https://ascopubs.org/doi/full/10.1200/JCO.2014.55.5730.

10. Domenico Ribatti, "Napoleone Ferrara and the Saga of Vascular Endothelial Growth Factor," *Endothelium* 15, no. 1 (Jan.–Feb. 2008), https://doi.org/10.1080 /10623320802092377.

11. Herbert Hurwitz et al., "Bevacizumab plus Irinotecan, Fluorouracil, and Leucovorin for Metastatic Colorectal Cancer," *New England Journal of Medicine* 350, no. 23 (2004), https://www.nejm.org/doi/full/10.1056/NEJMoa032691.

12. Line Ledsgaard et al., "Basics of Antibody Phage Display Technology," *Toxins* 10, no. 6 (2018), https://www.ncbi.nlm.nih.gov/pmc/articles/PMC6024766.

13. Rodrigo Barderas and Elena Benito-Peña, "The 2018 Nobel Prize in Chemistry: Phage Display of Peptides and Antibodies," *Analytical and Bioanalytical Chemistry* 411 (2019), https://link.springer.com/article/10.1007/s00216-019-01714-4.

14. André Frenzel, Thomas Schirrmann, and Michael Hust, "Phage Display–Derived Human Antibodies in Clinical Development and Therapy" *MAbs* (Oct. 2016), https://www.ncbi .nlm.nih.gov/pmc/articles/PMC5058633.

15. E. Dolk et al., "Isolation of Llama Antibody Fragments for Prevention of Dandruff by Phage Display in Shampoo," *Applied and Environmental Microbiology* 71, no. 1 (Jan. 2005), https://doi.org/10.1128/AEM.71.1.442-450.2005.

16. Shahir S. Rizk et al., "Targeted Rescue of Cancer-Associated IDH1 Mutant Activity Using an Engineered Synthetic Antibody," *Scientific Reports* 7, no. 1 (Apr. 3, 2017), https://doi.org /10.1038/s41598-017-00728-1.

17. Yinan Wei and Michael H. Hecht, "Enzyme-Like Proteins from an Unselected Library of Designed Amino Acid Sequences," *Protein Engineering, Design and Selection* 17, no. 1 (Jan. 2004), https://doi.org/10.1093/protein/gzh007.

18. Bassil I. Dahiyat and Stephen L. Mayo, "De novo Protein Design: Fully Automated Sequence Selection," *Science* 278, no. 5335 (1997), https://www.science.org/doi/10.1126 /science.278.5335.82.

19. "The Serious Search for Black Faculty at MIT," *Journal of Blacks in Higher Education* 7 (1995): 41–43, https://doi.org/10.2307/2963425.

20. Robert Bruce Slater, "The First Black Faculty Members at the Nation's Highest-Ranked Universities," *Journal of Blacks in Higher Education* 22 (1998): 97–106, https://doi.org/10 .2307/2998851.

21. Brian Kuhlman et al., "Design of a Novel Globular Protein Fold with Atomic-Level Accuracy," *Science* 302, no. 5649 (Nov. 21, 2003), https://www.science.org/doi/10.1126 /science.1089427.

22. D. Röthlisberger et al., "Kemp Elimination Catalysts by Computational Enzyme Design," *Nature* 453, no. 7192 (May 8, 2008), https://doi.org/10.1038/nature06879.

23. Lin Jiang et al., "De Novo Computational Design of Retro-Aldol Enzymes," *Science* 319, no. 5868 (Mar. 7, 2008), https://doi.org/10.1126/science.1152692.

24. Jiayi Dou et al., "De Novo Design of a Fluorescence-Activating β-barrel," *Nature* 561, no. 7724 (2018), https://pubmed.ncbi.nlm.nih.gov/30209393.

25. Brian Koepnick et al., "De Novo Protein Design by Citizen Scientists," *Nature* 570, no. 7761 (June 20, 2019), https://doi.org/10.1038/s41586-019-1274-4.

26. Rhiju Das et al., "Structure Prediction for CASP7 Targets Using Extensive All-Atom Refinement with Rosetta@home," *Proteins: Structure, Function, and Bioinformatics* 69, no. S8 (2007), https://www.researchgate.net/publication/5951040_Structure_prediction_for _CASP7_targets_using_extensive_all-atom_refinement_with_Rosettahome.

27. Robert Kleffner et al., "Foldit Standalone: A Video Game–Derived Protein Structure Manipulation Interface Using Rosetta," *Bioinformatics* 33, no. 17 (2017), https://www .researchgate.net/publication/316843310_Foldit_Standalone_a_video_game-derived _protein_structure_manipulation_interface_using_Rosetta.

28. Longxing Cao et al., "De Novo Design of Picomolar SARS-CoV-2 Miniprotein Inhibitors," *Science* 370, no. 6515 (Oct. 23, 2020), https://doi.org/10.1126/science.abd9909.

29. James Brett Case et al., "Ultrapotent Miniproteins Targeting the SARS-CoV-2 Receptor-Binding Domain Protect against Infection and Disease," *Cell Host & Microbe* 29, no. 7 (2021), https://www.sciencedirect.com/science/article/pii/S1931312821002869.

30. John Jumper et al., "Highly Accurate Protein Structure Prediction with AlphaFold," *Nature* 596, no. 7873 (Aug. 26, 2021), https://doi.org/10.1038/s41586-021-03819-2.

31. Robert F. Service, "'The Game Has Changed.' AI Triumphs at Protein Folding," *Science* 370, no. 6521 (Dec. 4, 2020), https://www.science.org/doi/10.1126/science.370.6521.1144.

32. Pierre-Emmanuel Y. N'Guetta, Maggie M. Fink, and Shahir S. Rizk, "Engineering a Fluorescence Biosensor for the Herbicide Glyphosate," *Protein Engineering Design & Selection* 33 (2020), https://academic.oup.com/peds/article/doi/10.1093/protein/gzaa021 /5905303.

33. Alessio Nocentini, Claudiu T. Supuran, and Clemente Capasso, "An Overview on the Recently Discovered Iota-Carbonic Anhydrases," *Journal of Enzyme Inhibition and Medicinal Chemistry* 36, no. 1 (2021), https://www.ncbi.nlm.nih.gov/pmc/articles/PMC8425729.

34. Oscar Alvizo et al., "Directed Evolution of an Ultrastable Carbonic Anhydrase for Highly Efficient Carbon Capture from Flue Gas," *Proceedings of the National Academy of Sciences USA* 111, no. 46 (Nov. 18, 2014), https://doi.org/10.1073/pnas.1411461111.

35. Sintawee Sulaiman et al., "Isolation of a Novel Cutinase Homolog with Polyethylene Terephthalate-Degrading Activity from Leaf Branch Compost by Using a Metagenomic Approach," *Applied and Environmental Microbiology* 78, no. 5 (2012), https://www.ncbi nlm .nih.gov/pmc/articles/PMC3294458.

36. V. Tournier et al., "An Engineered PET Depolymerase to Break Down and Recycle Plastic Bottles," *Nature* 580, no. 7802 (Apr. 2020), https://doi.org/10.1038/s41586-020-2149-4.

37. Robert Grace, "Carbios Gets Green Light to Build a PET Bio-Recycling Plant in France," *Plastics Engineering* (2023), https://www.plasticsengineering.org/2023/10/carbios-gets -green-light-to-build-a-pet-bio-recycling-plant-in-france-002488.

ACKNOWLEDGMENTS

THANK YOU TO our agent, Luba Ostashevsky, at Ayesha Pande Literary Agency and our editor, Rachel Field, not only for their help and support during this entire process, but also for believing in the power of art and science. Thank you to everyone at Harvard University Press who helped bring this book together. Thank you to Professor Kelcey Ervick for the inspiration to practice the habit of making art. Thank you to the Research Corporation for Science Advancement for their support through the Cottrell Scholar Award. Thank you to everyone at Burroughs Wellcome Fund for giving us the time and space to write and illustrate. And a special thanks to our family and friends for their support and encouragement as we pursued our goal of writing this book.

INDEX